CONTENTS

PREFACE

The original proposal for this book was submitted in April 1990 and claimed that the final manuscript, then titled 'Space for desire', would be delivered in summer 1992. So, I have had some time to think about the cover (never mind the contents) for *The Body and the City.* I collected many postcards, but none seemed quite right. About this time last year, though, I went to an exhibition at the Saatchi Gallery in London. My new colleagues at the Open University were being (and stayed) exceptionally sympathetic to, and supportive of, my wish to write this book, and – although I had been telling Tristan Palmer (the editor at Routledge, whose tolerance and patience cannot go without thanks) for years that it was almost done – it now seemed possible that I would actually finish it. So, I was wandering around the gallery and generally enjoying the collection, but with half a mind on a possible cover photograph, when I came across Roger Mayne's 'Man and Shop Window, Rue de Reaumur, Paris'.

As I looked at the photograph, my first reaction was to wonder whether the mannequins in the shop window were in fact real people. Then it crossed my mind that this photograph might be a set up (where actors play dummies so that the sly or shocked reactions of passers-by can be caught forever), along the lines of Robert Doisneau's *Un Regard Oblique* series. I stepped closer to the photograph, but I still could not work out whether the bodies were people or mannequins. Perhaps this image was much more like the work of Helmut Newton. Either way, it dawned on me that I could be scrutinising a voyeuristic or titillating photograph of naked women (and one man). I felt embarrassed and ashamed. I was still intrigued by the image's ambiguity, but also increasingly uncomfortable with, and disturbed by, my first reaction. Partly, then, it is this moment of (my) fantasy, curiosity, ambivalence and embarrassment that makes this an appropriate image for the cover.

There are other (probably better) reasons. There seems to be, here, a story about power (or at least freedom). The man walking the streets is unaware, either of the photographer, or of the bodies in the shop window. He seems free, in this space, to ignore the world around: power also resides in the taken-for-granted, in the freedom not to be aware of the world around – power does not just reside in the surveillance of, and overt control over, the world. The image unblinkingly depicts both these moments of power.

Meanwhile, the mannequins tell of the stripping of the body down to idealised parts, of the role of the part-body in commodification, of the display of commodities through the lens of the shop window, and of the commodification of the streets. Furthermore, this is a City scene, the photograph shouts PARIS, while the street, the shop windows, the pavement, the parked car, the business man, the mannequins become the frozen elements of a specifically urban story: commerce, commodities, circulation, signs, writing on the walls, across organised and unruly space. Moreover, the graffiti tags and the fly-poster, on the shop-front shutter, mark the transmission of other signs – perhaps this is the subversive assertion of alternative identities, or maybe something more prosaic. The image shows that the City traffics in bodies, meanings and modalities of power.

In this image, there are bodies in a city, so at this level this is an 'obvious' choice for the cover. However, the most important factor was the moment of ambiguity and ambivalence, the simultaneous fixity and fluidity in the scene. I could not make out the bodies in the shop window. While other elements were apparently obvious, these seem to be opposing things: Mannequins or People? Then, again, they were also the same thing: Models. The meaning of the picture shifted around these fixed and fluid co-ordinates. As a result, the scene, the City, became different things: an immutable site of dead signs, or a living part of everyday life, or the vital place of an erotic urban unconscious, or the abstract space of unyielding capitalist social relations? Whatever, the Models (Mannequins, People) are 'fantastic', points of capture for (my) fantasy, (my) desire and (my) power. For me, Mannequins or People, these fixed bodies oscillated between meanings – never settling down into a cosy 'truth'. I was constantly located, dislocated and relocated by the tensions between my senses, thoughts and feelings about the scene. And none of this took place in the abstract – me 'versus' the image – there are other spatialities: the friend I went with (who doesn't get on with this photograph), the gallery, North London, the book, and so on.

Maybe you felt or thought something like this when you saw the cover of *The Body and the City* – I have shown the picture to others and they (men and women) have reacted in different ways: some thought that the bodies might be 'dummies', others felt they were 'actors', and a few seemed to think it was all too obvious. More recently, I showed the image to someone who reacted by saying that I couldn't use this on the cover, that it was another laddish image, that she was sick of seeing heterosexual men's desire, that she didn't want to have my desires or fantasies imposed on her, and that using this image was a sign of my control, my power. I was deeply upset: first, because I don't want to make women angry, and especially not her; and, second, because I once again felt ashamed by my first impressions – I should have felt angry too. We are – as friends – still discussing this image and the questions it foregrounds, and forecloses, about the body, meaning and power relations, but this exchange emphasises that what is at stake in power-laden sexualised and gendered 'ways of seeing' is the status of the female body. This is not a new point, but questions remain on the specific ways that regimes of the body and the spaces of subjectivity intertwine. Lines of desire, disgust and power radiate out both from the bodies in the photograph and from the eyes of the viewer, building intricate patterns of meaning, identity and power relations, across space and through space. And it is at this point that it is possible to suggest that a simultaneously geographical and psychoanalytic imagination will say something significant about the body and the city.

Steve Pile (London and Milton Keynes)
September 1995

ACKNOWLEDGEMENTS

The author and publisher would like to thank the following for permission to reproduce material:

Famous Blaggers, No. 83, © BIFF Products;

Sur la Tour Eiffel, Paris 1938 by Erwin Blumenfeld, © 1985 Éditions du Désastre; by permission of Kathleen Blumenfeld;

Montmartre, 1950 by Édouard Boubat, © 1988 TOP/Éditions du Désastre; by permission of Agence TOP, Paris;

Früh um 5 Uhr! 1920–21 by George Grosz, © DACS 1996;

'The Knight's Move' series 1990 by Tessa Boffin, in Tessa Boffin and Jean Fraser (eds) *Stolen Glances: Lesbians Take Photographs*, Pandora Press.

INTRODUCTION

1

INTRODUCTION
the unknown spaces of the mind

BANNER HEADLINE, *Daily Mail*, 22 SEPTEMBER 1995

EURO COURT'S GIPSY SHOCK

It looks like the *Daily Mail* has been hit by lightning today. First paragraph:

A SHOCK European ruling threatens to throw Britain's planning laws into chaos

(front page).

Is this a familiar Euroscare along the lines of 'they' want to ban 'our' sausages or Babycham or the like? It is a good front page for a patriotic (and xenophobic) tabloid like the *Daily Mail*: hundreds of thousands of middle Englanders are quaking in their boots at the thought of what might follow. There is a threat to the state and good order from a familiar menace: the European Court of Human Rights. Worse, for them, is to follow. Next paragraph, the *Mail* has a prophecy: 'It could mean gipsies being allowed to park their caravans virtually anywhere they choose, even in rural beauty spots where houses would never be allowed.'

On page 2, the newspaper of the year 'exposes' the issues. The case was brought to the Court by Mrs June Buckley after she was refused planning permission by her local council. The *Daily Mail* decides to side with 'the public interest', rather than the 'slight' right of Mrs Buckley to her lifestyle. To support its case, the *Mail* turns to Kathleen Crandall, South Cambridgeshire council's legal and housing director, to find out what she thinks. She finds caravans 'intrusive' and they 'deface' the landscape. She warns, 'if this [judgement] goes against us, it would mean everywhere is affected'.

Back to the front page. Apparently, it's an 'alarming prospect' and, allegedly, it has 'provoked fury' amongst (unnamed) Tory MPs. The *Daily Mail* alleges that the ruling by the Court of Human Rights states that 'laws requiring planning permission violate gipsy rights to "a traditional life-style"'. Although nothing has been finalised, this decision is a 'bombshell' which has led to calls from many (undisclosed) quarters to 'pull out' of the European Court. Sir Ivan Lawrence, QC, is quoted as saying that this is one more example of 'interference in our sovereign democratic nation'. It gets worse. Five column inches into the 'shocking' front page story: 'Furious Whitehall officials fear the case could also open the door to thousands of non-gipsies putting mobile homes in fields and gardens for their parents or children.' The *Daily Mail* is spurred to make its own comment (on page 8). Once more, it complains, Britain has faced 'foreign judges in an alien court' and 'lost'. It may seem like 'small beer', but this 'judgement' could lead to 'a flood of people' putting 'their caravans anywhere they like'. The rhetoric builds to a climax:

> It may be time to change our polite, gentlemanly approach. Bad enough that we regularly have to submit to that other European Court . . . But when we also feel obliged to swallow the most crass decisions of the entirely separate Court of Human Rights, the limit has been reached . . . the Government should insist on judgement in the proper place. Before the British courts.
>
> (*Daily Mail*, 22 September, page 8)

Meanwhile on page 5, the newspaper of the year carries another gloomy story. It is a '**Tale of two families**', called 'the Smiths' and 'the Monks'. The two families are on opposite sides of a dispute between a private estate and an estate currently under construction by a housing association. The two estates, apparently, are 'a stone's throw apart'. The owners of the private houses anticipate that the value of their homes will decline by at least 10 per cent when their new neighbours move in. They have a solution to this 'problem': they want the council to build a wall between the two estates! The fearless *Daily Mail* has sought the human story behind this conflict, as the 'wives go to war': Tina Smith has led the campaign for 'segregation', while Tina Monk is amongst the 63 potential tenants. The women have already been in doorstep confrontation, which – the *Daily Mail* sounds disappointed to find – was tense, but peaceful.[1]

There isn't really a full page story here, but the newspaper of the year pads it out with some interesting comparisons. The Smith husband earns £18,000 a year, while the Monk family have an annual income of £10,500 – neither are particularly rich, nor especially poor. Money is tight for both families. Hardly grounds for the 'fear' that appears to have been provoked by these

---

1 Suspiciously, this only took place the day before publication: surely it is not possible that the *Mail* set the 'confrontation' up?

allegedly 'undesirable poor'. The real problem, for Tina Smith, seems to be that families like the Monks, who have eight children and 'may have more babies', will be moving into the nearby estate. It is the presence of 'poor' children, within 'a stone's throw', which is actually going to be undesirable. In the end, Tina Smith says that she isn't 'a snob', all she wants is to 'protect the value of my property' (which her husband works all hours to pay for), while Tina Monk thinks that 'all this fuss is silly'; she concludes that 'if they want to put up a wall, that's their business'.

These two stories are of quite different orders,[2] but they help illustrate what is at stake in juxtaposing a geographical and a psychoanalytic imagination. At one level, these stories are easy to interpret: the first tells of racism (the 'English Gent' versus 'Johnny Foreigner'), while the second is about the tension between middle-class and working-class interests (the Smiths versus the Monks). So far, neither geography nor psychoanalysis is essential to these interpretations. On the other hand, these stories tell of border disputes and of the 'shock', 'fear' and 'fury' that the transgression of borders provokes. These are stories about the intertwining of territories and feelings, about demonised others, and about senses of self and space. It is clear to me that these situations cannot be fully understood without a geographical and a psychoanalytic appreciation of the 'psycho-spatial' dynamics. Let me briefly tease out some ideas, which I think will introduce the substantive concerns of this book.

The European ruling that people can set up homes on their own property ought to be something the *Daily Mail* supports (they have in the past relished campaigns against the 'little Hitlers' in 'loony left' councils). Nevertheless, it is a 'shock', a shock which has two sources. The first is the clash between the space of the sovereign British nation-state and the space of the European Union and its analogues. The battle lines of this 'invisible' territorial dispute are familiar: 'we' are polite and gentlemanly, 'they' are alien, foreign, out of touch and wield arbitrary power over 'us', to which 'they' have no right. The grounds of this dispute are inherently, and obviously, spatial as well as social. The second source of the shock is just as obvious, but also less so. The banner headline screams 'GIPSY'. Invidious and disingenuous racism is used to prop up and invigorate xenophobia. Gypsies rarely own the property they live on, and they live and work predominantly in urban areas,[3] and are therefore almost wholly excluded, first, from the recommended judgement and, second, from the alleged potential for the despoliation of rural beauty sites.

There is more to this story, though: not only does the narrative produce, mix and circulate racist signifiers, it also infuses them with the intemperate feelings of never named people. Whether these feelings exist or not is

2 Indeed, neither are 'stories' since the European Court of Human Rights has not made a decision, while the new estate has not been built yet. They are more fantasy than 'fact', more hysteria than investigative journalism, which I think adds all the more to the case for a psychoanalytic reading.

3 See the Introduction to Part II, pp. 88–91.

irrelevant, the newspaper of the year makes it appear as if these violations of space (by Europeans, by gypsies) are personally felt as violations of the self. Civil servants, lawyers, Tory MPs 'fear': they are 'furious', threatened, alarmed, and so on. The *Mail* generates an economy of subjectivity and space which is then sold to the reader by alleging that these incursions into the nation, across borders, into the body could happen 'anywhere', as metaphors like 'bombshell', 'open the door', 'a flood of people' anticipate the disintegration of life as the reader knows it. The *Daily Mail* emphasises and circulates feelings at an unconscious, as well as conscious, level. In this story, the geographical and the psychoanalytic are ever present.

Meanwhile, the idea of the wall between 'the Smiths' and 'the Monks' makes concrete both the significance of border disputes and the sense of violation of self which comes with the wrong people turning up in the wrong place. The wall is an armour against the other – it is meant to shore up those feelings of 'fear' and 'disgust': a hard, high, fixed, impermeable boundary on a space which is both urban and bodily. The wall is not yet real, but it is very real: a focus and sign of the exasperation of a group of people. It cements associations between their situation and with Berlin, and maybe Northern Ireland, with other states of terror. Yet, from afar, this terror seems to have no substance – not only do housing association tenants commonly own a proportion of their property, but they also need to convince the housing association of the security of their income. They are not the poorest of the poor, nor are they some kind of unruly underclass. Yet it is to these apparently irrational connotations – mobilised through the idea that these families have 'eight children', living only 'a stone's throw' away – that 'the Smiths' are responding. The reality of this dispute is clearly about the definition and control over class borders, but there are emotional investments here which go way beyond the cash price of a house.

What is at stake here is an appreciation of the intricate and dynamic ways in which narratives of space and self intertwine. These stories are about the ways in which people gain a sense of who they are, the ways in which space helps tell people their place in the world, and the different places that people are meant to be in the world. So, in the newspaper articles, there are underlying senses of people having a 'proper place' and of people who are 'out of place'. These are simultaneously geographical and psychoanalytic tales, so in this book I have attempted to write a dialogue between, and within, both discourses. I should set the scene for this narrative, first, by providing a thumbnail sketch of psychoanalysis and, then, by thinking about geography and the mind.

THE *TERRAE INCOGNITAE* OF PSYCHOANALYSIS

Historically, psychoanalysis is the term given to a system of thought, which was created by Sigmund Freud. Born in 1856, Freud was the son of a moderately successful Jewish wool merchant, living in the Austro-Hungarian Empire. Originally, Freud trained as a medical doctor, with the ambition of alleviating suffering. In 1886, Freud travelled to Paris to study neurology with Jean-Martin Charcot. This experience led Freud to consider the problem of hysteria: hysterics had bodily symptoms, such as paralysis or lumps in the throat, for which no anatomical cause could be found and which disappeared when the hysteric was hypnotised.[4] Freud's initial attempts to cure hysteria and the subsequent failure of his cure, despite some ostensible success,[5] led him over many years to propose a new system of thought: psychoanalysis. Freud came to believe that the patient's internal psychic conflict was being symbolically expressed in bodily symptoms – this understanding of the mind and the body remain at the centre of psychoanalysis. Indeed, the existence of an unconscious, the vicissitudes of desire, the secret life of things (for example, phallus-shaped) and the slippages of meaning, and the hiding and inadvertent expression of innermost feelings are often implicit in our common-sense understandings of ourselves. However, Freud's writings are still controversial and there is nothing in psychoanalysis that is generally accepted.[6] Most importantly, there has been a tremendous amount of speculation about the realm of the unconscious and the so-called sexual instincts.

Arguably, the most fundamental concern of psychoanalysis is with the existence of a dynamic unconscious. It is by revealing the forces operating in the unconscious that human behaviour (individual and/or group) can be understood. Crudely, the unconscious is an area of psychological functioning that is not accessible to the subject, but which nevertheless has a motivating influence on their everyday lives: their thoughts, feelings and actions. However, the structure and content of the unconscious are a matter of some considerable debate.[7] For most analysts, however, the unconscious is made up of the residues of infantile experiences and the representatives of the person's (particularly sexual) drives. Although there is considerable disagreement about how children develop increasingly intricate and dynamic psychological structures, the experiences of early childhood are generally accepted to be critical. Basically, it is argued that the child develops defences against painful experiences, mainly by keeping them away from consciousness and, commonly, by hiding them in the unconscious: this is called repression.

The effect of repression is to produce an internal splitting of the mind into a conscious and an unconscious. The unconscious is not static, but has its own dynamics. Most importantly, while the unconscious does not determine what goes on in the mind, it continually seeks to find expression by fighting a kind of guerrilla war with the conscious: this is most vividly experienced

---

4 Much can be made from the fact that these hysterics were predominantly women.
5 The patient's symptoms disappeared under hypnotic suggestion, only to reappear in another form later.
6 From the perspective of social theory, introductions to psychoanalysis can be found in Bocock, 1976; Frosh, 1987; Craib, 1989 and Elliott, 1992, 1994.
7 Compare, for example, Irigaray, 1974 or Zizek, 1989.

in dreams. It is the unconscious that is responsible for producing feelings, thoughts and actions, which cannot be readily explained by the person experiencing them. Freud's conception of the unconscious means that consciousness cannot form the basis for understanding human behaviour and experience. People's choices are motivated and constrained by forces that lie outside their control or easy access.

Moreover, Freud provides a developmental account of the psyche that simultaneously reveals the ways in which people give meaning to their world (of people, events and things) and receive meaning from that world, where they act according to the interactions between these worlds, and where people are resourceful and devious in the ways that they deal with, and express, the pleasures and pains that they live through. Other important aspects of Freud's work include a theory of dreams and a methodology for interpreting this and other psychic phenomena (such as the infamous Freudian slip), a theory of instincts, and other models of the mind.[8] Working through the inconsistencies, problems and unacceptable aspects of Freud's thinking has meant that psychoanalysis has developed many lively schools of thought – varying from ego psychology to Lacanian psychoanalysis, from Jungian psychology to feminist reinterpretations of psychoanalytic precepts.

There is another side to this heated debate: Freud's ideas have been shown to have an implicit moral scheme, which has rightly attracted much criticism and anger. Foucault and Irigaray see psychoanalysis itself as a form of repression, as a micro or macro tactic of power (see Foucault, 1961, 1966; Irigaray, 1974, 1977). Meanwhile, certain feminists have argued that psychotherapy replicates the situation of father–daughter rape (Ward, 1984). From a Marxist perspective, Timpanaro (1976) argues that psychoanalysis is incapable of seeing beyond the ideological level to class interests, while others have argued that it systematically disguises social structures, depersonalises the individual and privatises distress (Brooks, 1973). From this perspective, psychoanalysis is a bourgeois discipline, through which the bourgeoisie confirm their decadence and moral bankruptcy. In this sense, psychoanalysts act as capitalism's psychic first-aiders. Psychoanalytic discourse, then, is highly contested – both within its borders and from outside. So, any attempt to read the relationship between the subject, space and the social using psychoanalysis must be wary that the letter could be a bomb. And, any encounter between a geographical and a psychoanalytic imagination must be partial and selective (neither discourse is free of corrupt connotations).

More hopefully, much contemporary psychoanalysis is far less concerned with Freud's endeavour to provide a scientific and/or universal account of the human psyche, than to account for the personal meanings that people produce for themselves as they struggle to cope with, and make sense of, the

8 A summary of Freud's life and work can be found in Gay, 1988 and Wollheim, 1991 (but see also Ricoeur, 1970). On Freud's topologies of the mind, see Chapter 4 below.

painful realities of everyday life. Partly because of its internal disputes and partly despite them, psychoanalysis suggests both sensitive clinical (therapeutic) practices and insightful interpretations of the relationships between people, other people and their worlds. Furthermore, psychoanalysis and geography share an interest: it has long been a concern of geographers to discover the *terrae incognitae* of people's hearts and minds. Initially, this exploration involved a discussion of the geographical imagination. A brief description of two articles shows just how important notions of fantasy, desire, the body and psychological development were to some geographers. Despite crossing psychoanalytic concerns, however, psychoanalysis played only a supporting role in these explorations.

TERRAE INCOGNITAE: 'THE HEARTS AND MINDS OF MEN'

In 1947, J. K. Wright boldly stated that there were no more lands left for geographers to go out to explore. Although he believed that nowhere on earth had not already been trampled over by somebody, he also argued that even 'if there is no *terra incognita* today in an absolute sense, so also no *terra* is absolutely *cognita*' (page 4). While the geographer has no place left to explore or chart, there is still much to do. He (for his is a specifically masculine imagination) must concentrate on a region and explore the boundless obscurities that lie within it: 'the unknown stimulates the imagination to conjure up mental images of what to look for within it, and the more there is found, the more the imagination suggests for further search' (page 4). The need for some kind of geographical knowledge was, he presumed, 'universal among men', though 'its acquisition, in turn, is conditioned by the complex interplay of cultural and psychological factors' (page 14). It is this relationship between the universal need for geographical knowledge and the complex interplay of cultural and psychological factors that will eventually preoccupy Wright – and many other geographers in the ensuing years.

There is at the heart of Wright's geographical imagination a mythic sense of awe and a sense of male heroism: the geographer's desire is likened to Odysseus's desire to hear the Siren's call.[9] For Wright, this calling is unique to each individual, and demands a philosophy of geography which is sensitive to the perspective of each individual. Intriguingly, Wright comments:

> A great deal had been written and more said about the nature of geography; far less about the nature of geographers. Could we subject a few representative colleagues to a geographical psychoanalysis, I feel sure that it would often disclose the geographical libido as consisting

9 See Gillian Rose's analysis of the gendering of these kinds of metaphors (1993, Chapter 5).

fully as much in aesthetic sensitivity to the impressions of mountains, desert, or city as in an intellectual desire to solve objectively the problems that such environments present.

(Wright, 1947: 9)

Wright psychoanalyses the libido (or sexual energy)[10] of geographers and discovers that it has two sides: it is both an aesthetic sensitivity and an intellectual desire to solve problems. The universal nature of geography consists in a double appreciation, first, of the beauty of the landscape or cityscape and, second, of the need to solve problems. The geographical libido lies at the heart of his desire to extend geography into new *terrae incognitae*: on the one hand, the not fully *terrae incognitae* of the external world of mountains, deserts and cities; and, on the other, the undiscovered *terrae incognitae* of the internal world of desire, fear, fascination, illusion, error, greed, prejudice, partiality, intuition and imagination. Wright closes his paper with this stirring conclusion: 'the most fascinating *terrae incognitae* of all are those that lie within the minds and hearts of men' (Wright, 1947: 15).

It is exactly here, over a decade later, that David Lowenthal opens his discussion of 'geography, experience and imagination' (1961). Unlike Wright, he is not interested in psychoanalysing geographers' libido. On the other hand, he is open to psychoanalytic findings where they support his case that people's hearts and minds contain the most fascinating *terrae incognitae* of all (Lowenthal, 1961: 241). Lowenthal's paper is a wide and erudite exploration of the ways in which people experience and imagine space and time. Lowenthal draws on anthropological evidence to back his assertion that there is a physical realm, which is common to all people who experience it, even though that world can be symbolised in many different ways. This external world is not just a world of co-existing facts, it is also profoundly spatial.

Lowenthal begins by analysing the ways in which individuals understand their external worlds. Although understanding is basically universal, he finds significant limitations to some people's spatial abilities: 'The most fundamental attributes of our shared view of the world are confined, moreover, to sane, hale, sentient adults. Idiots cannot suitably conceive space, time, or causality. Psychotics distinguish poorly between themselves and the outside world' (Lowenthal, 1961: 244). There are wildly differing world-views; claustrophobics do not experience the world in the same way to mystics, for example, while the mad, children, and the unfit cannot adequately conceive space, time and causality. Like Wright, Lowenthal makes an analytical distinction between the internal world and the external world. To back up

10 In 1905, Freud suggested that libido was a physical sexual energy, which was 'invariably and necessarily' masculine in nature (pages 138–140 and 141). The presumption that libido is masculine *and* male underlies and underpins Wright's analysis. However, Freud refused to correlate the categories of sex ('male' and 'female') with gender ('masculine' and 'feminine') (see 1905: 141–144).

this assertion, Lowenthal draws on the work of psychoanalysts. Here, he uses the work of Money-Kyrle on 'the world of the unconscious and the world of common sense' as evidence. Lowenthal cites the effects of illness and injury on both perception and cognition, referring to the work of Fenichel on 'the psychoanalytic theory of neurosis' (and various other papers from mainstream psychology). Moreover, Lowenthal suggests that 'to see the world more or less as others see it, one must above all grow up; the very young, like the very ill, are unable to discern adequately what is themselves and what is not' (ibid.: 244).

The fundamental attributes of the universal need for geographical knowledge are connected to being grown-up, healthy and normal. It seems that the interest, here, is not in the hearts and minds of 'men', but in being sure that geography is a sane, hale and adult world-view. This issue of a grown-up world-view leads Lowenthal to Jean Piaget's discussion of the psychological development of children. Lowenthal asserts this: 'unable to organize objects in space, to envisage places out of sight, or to generalize from perceptual experience, young children are especially poor geographers' (ibid.: 245).

The point of these examples is to stress that personal concerns are unique, but that they are also learnt. Lowenthal once again cites the psychoanalytic literature on the child in its maturational environment, drawing on the work of Erik Erikson (1950). We are all the captives of our own personal histories. Moreover, this is not just confined to childhood. So, Lowenthal also uses the work of Kevin Lynch (1960) to suggest that people employ different landscape features as points of reference. Importantly, Lynch also demonstrated the personal elements of personal geographies: 'all information is inspired, edited and distorted by feeling' (Lowenthal, 1961: 257). Though these are in some senses uniquely personal geographies, there are also regularities shared by groups: 'territoriality – the ownership, division, and evaluation of space – also differs from group to group' (ibid.: 253). And this is organised through language, which organises the world, just as it organises the words which seemingly describe the world. Thus, Lowenthal cites Friedrich Waismann as saying that

> 'by growing up in a certain language, by thinking in its semantic and syntactical grooves, we acquire a certain more or less uniform outlook on the world ... Language shapes and fashions the frame in which experience is set, and different languages achieve this in different ways.'
>
> (Waismann, 1950–1; cited by Lowenthal, 1961: 255)

On the theme of the limits to spatial abilities, Lowenthal notes that there are differences between the ways in which men and women perceive the environment, noting that, according to Witkin (1959), 'strong-minded men are better at telling which way is up than women, neurotics, and children' (page 256). He also notes that the ordering of gendered language is not so clear cut: the sun is masculine in French, but feminine in German; the moon is masculine in German, but feminine in French. Culture orders what Nature provides. Into this melting pot, and along the lines of culture and language, Lowenthal adds evidence derived from Margaret Mead to describe the perceptual imagination of so-called primitive peoples. On this evidence, he believes that just as children grow up and become better and better geographers, so do civilisations: 'The shared world view is also transient ... every generation finds new facts and invents new concepts to deal with them' (Lowenthal, 1961: 243).

Significantly, when the world-views are analysed by Lowenthal, a series of inferior geographers are located: children, the mad, the superstitious, primitives and women. In these at least racist and misogynist terms, the past of an adult, rational male and Western geography is to be found in its others, but not in itself. Geography has left behind its more primitive, childlike, feminised world-views forever: after all, Lowenthal reminds us, as Heraclitus said, you can't step into the same river twice.

Lowenthal is quick to point out that (all) people's world-views are partial – and they are (all) centred on people's (shared) experience of their bodies. Nevertheless, the suspicion remains that some people are more partial than others. Moreover, partiality is measured against the standard of natural human interests, such that 'purpose apart, physical and biological circumstances restrict human perception' (page 246). But Lowenthal now shifts the discussion of these circumstances onto the terrain of a generalised body. Thus, everybody's body has particular capacities: for example, they see light in a particular spectrum, and sense heat in specific ways. Again the psychological and philosophical literature are cited to establish his case that human bodies are real limits to people's perception. The question lingers as to how certain differences are to be explained if personal geographies are personal and the body is a universal norm. Having placed people at the centre of their world, and having located them in their limited fragile bodies, Lowenthal is in a position to suggest that people create personal geographies, which are 'separate personal worlds of experience, learning, and imagination [which] necessarily underlie any universe of discourse' (page 248).

People do not just exist in their own little worlds, however, they learn a shared picture of the world. Indeed, a sense of a solid shared world and a stable sense of ourselves within that world is seen to be essential for our psychic and physical survival – once more drawing on the psychoanalytic work of Money-Kyrle (1956: 96). These secure personal geographies are localised and restricted, while it is the shared public geographies which transcend objective reality. Nevertheless, each of the infinite number of private geographies includes elements that are not found in the shared world-view, partly because this general world-view is the result of the consensual universe of discourse. While the communal world-view includes the hopes and fears of a culture, 'fantasy plays a more prominent role in any private milieu than in the general geography. Every aspect of the public image is conscious and communicable, whereas many of our private impressions are inchoate, diffuse, irrational, and can hardly be formulated even to ourselves' (Lowenthal, 1961: 249).

So far, Lowenthal has built up a series of analytical divisions: between personal and general geographies, between children's and adult's geographies, between the irrational and the rational, between strong-minded men and women, between the real world and the world of fantasy, between the unconscious and the conscious, between the mad and the sane. These pairings are a series of bi-polar opposites: they only make sense in relation to one another and they only make sense if they are quite unlike each other. Thus, diffuse personal geographies are quite unlike communicable public images: the upsetting thing, for this view, is that 'fantasy' cannot be confined to one side of the fence or the other – what, for example, of religion? An irrational shared fantasy? In order to maintain the dualistic analysis, rational shared public images have to be cleansed of superstition, while individuals must remain resolutely dreamy:[11] 'Hell and the garden of Eden may have vanished from most of our mental maps, but imagination, distortion, and ignorance still embroider our private landscapes' (page 249).

As society grows up, it loses its superstitions, even while individuals still cling to their irrational worlds.[12] Unsurprisingly perhaps, Lowenthal finds evidence for the irrationality of an individual's mental maps in psychoanalytic concerns, such as dreams (from Money-Kyrle, 1960: 171). Indeed, Lowenthal begins to build quite a complex relationship between the mind, geographical elements such as landscape and the outside world, which is always fictional in some senses. From this perspective, 'What people perceive always pertains to the shared "real" world; even the landscapes of dreams come from actual scenes recently viewed or recalled from memory, consciously or otherwise, however much they may be distorted or transformed' (Lowenthal, 1961: 249).

While society may be more and more grown up and rational, the world which people inhabit is still a world of dreams. Waking 'Reality' cannot shake off its Dreams. Moreover, the 'real' world is a sequence of memories, and dreams, conscious or otherwise. Lowenthal uses the work of psychoanalysts such as Fisher (1956), Knapp (1956) and Fisher and Paul (1959) to suggest that dreams, images and perception are linked in the mind, though not through a simple one-to-one correspondence. What Lowenthal implies is that there is an area beyond conscious thought which also conditions people's view of reality, but that it does so in a distorted way. In psychoanalytic terms, this is the unconscious, but Lowenthal does not pursue this understanding of mental life.

Under psychoanalytic pressure, the dichotomies have broken down: dreams inspire the 'real' world, while unconscious processes underlie the personal, the social and the spatial. Furthermore, fantasy is simultaneously social, spatial and personal. From here, Lowenthal could not only disrupt all his analytical bi-polar opposites, but also use psychoanalysis to suggest (a)

---

11 On dualistic thinking in human geography, see G. Rose, 1993: 66–78.
12 This attitude to religion is somewhat reminiscent of Freud's (1927).

why there is such an investment in them, (b) how they serve to define, maintain and reproduce relationships between self and other, (c) what the qualities of the relationships between the dualisms are, and most importantly (d) what might lie beyond them. He could, from here, traffic in geography and psychoanalysis. Instead, he turns to notions derived from Kurt Koffka's gestalt psychology (as would many others, see Chapter 2 below). From this perspective, 'essential perception of the world, in short, embraces every way of looking at it: conscious and unconscious, blurred and distinct, objective and subjective, inadvertent and deliberate, literal and schematic' (Lowenthal, 1961: 251).

Lowenthal has re-established the analytical grid of bi-polar opposites. The point of his argument is, however, to establish the importance of each individual's mental maps, that is their perception and cognition of the world. Gestalt psychology offers a set of parameters that can be used to understand the way the mind builds up spatial patterns which reflect the physical world. In this way, he hopes to set a new agenda for geography, where geographers would explore the *terrae incognitae* of the simultaneously real, experiential and imaginary, and profoundly personal, world. The individual's personal geography is indispensable, for Lowenthal, because it lies at the intersection of their image of the environment, their perception of the world, the general world-view and their life history. Following the work of Boulding (1956) and Money-Kyrle (1960), Lowenthal argues that,

> every image and idea about the world is compounded, then, of personal experience, learning, imagination, and memory. The places that we live in, those we visit and travel through, the worlds we read about and see in works of art, and the realms of imagination and fantasy each contribute to our images of nature and man.
>
> (Lowenthal, 1961: 260)

And these images and ideas about the world, Lowenthal argues, are profoundly connected to people's behaviour, which lies at the intersection of personal geographies and the general world-view; in this way, the human world is unified by people's judgements about aesthetics and beauty and the rational judgements of culture.[13] In putting the individual at the centre of geographical concerns, Lowenthal was challenging some dominant figures in Geography. For example, in an influential paper on historical geography, Carl Ortwin Sauer stated that 'human geography ... unlike psychology and history, is a science that has nothing to do with individuals but only with human institutions, or cultures' (Sauer, 1941: 7). Nevertheless, Sauer concluded that, as geographers,

> we deal not with Culture, but with cultures, except in so far as we

---

13 This reminds me both of Wright's geographer's libido which has an aesthetic sensitivity and a problem-solving side (see above) and of G. Rose's suggestion that masculinity in geography exhibits an aesthetic and a social-scientific side (1993: 10). See also footnote 10 above.

> delude ourselves into thinking the world over in our own image. In this great inquiry into cultural experiences, behaviors, and drives, the geographer should have a significant role. He alone has been seriously interested in what has been called the filling of the space of the earth with the wishes of man, or the cultural landscape.
>
> (Sauer, 1941: 24)

Lowenthal's work marks the beginning of a sustained inquiry by geographers into the relationship between the psychology of the individual (against Sauer's advice), their perception of the external world and their behaviour in that perceived world. This paper, as Sauer's before, contains hints of what might be achieved by a psychoanalytically-informed exploration of the *terrae incognitae* of people's hearts and minds. Sauer talks of drives and wishes filling the space of the earth (a male geographer's libido), while Lowenthal speaks of fantasy and dreams constituting the world. Both are interested in the relationship between the individual and culture.

It is Lowenthal, however, who recognises the importance of childhood development, of the body, of personal histories – and of personal geographies. Nevertheless, there are – as with Freudian psychoanalysis – undisclosed value systems which any geography of experience and imagination would have to step over. For me, Lowenthal's paper raises significant issues, not least of which is the role of fantasy in everyday life, but he does not develop the implications of presuming an unconscious life. For this reason, the *terrae incognitae* of the unconscious remains just beyond the horizon, and the possibility of elucidating a 'psychoanalysis of space' or a 'psychodynamics of place' is foreclosed. I want to open up this possibility, especially with respect to the relationship between the subject, society and space – in the body and in the city.

THE *TERRAE INCOGNITAE* OF *THE BODY AND THE CITY*

In order to set up a dialogue between geography and psychoanalysis, I have first tried to outline the ways in which the relationship between the subject, society and space have been understood, first by geographers and later by Freud and Lacan. I have not tried to review the enormous literatures involved, nor have I wanted to simply repeat people's arguments. Instead, I have developed my arguments out of the work discussed. In this sense, at least, this book is a thesis. Nevertheless, I have tried to make it possible to read the parts separately and to ensure that the chapters make sense on their own. Parts II and III have their own introductions, so I will say only a little about them here, but I should say something about Part I.

Lowenthal's article, described above, is a remarkable piece for another reason: it mixes two strands of thinking about 'personal geographies' that were later to become competing approaches to human geography. Evidence is used from psychology, psychoanalysis, anthropology, gestalt psychology, and phenomenology to weave an elaborate story about the relationship between geography, experience and the imagination. Many geographers, at this time, were becoming increasingly doubtful about the ability of the then dominant spatial sciences to explain spatial patterns. They, like Lowenthal, searched for alternative models of 'man'. This story is told in the first chapter of Part I, 'Environment, behaviour, mind'. Although behavioural geography initially succeeded in establishing, as Lowenthal wanted, the importance of considering the personal geographies, or mental maps, of individuals, it became clear that it too suffered from flaws.

Reactions against behavioural geography came from two sources: humanistic and radical geographers. They posited their own senses of what people are like, how they relate to the world, and how they (can) behave in the world. The development of these positions is charted in Chapter 3, 'Geographies of human agency'. Commonly, when the story of these developments is told in geography, people tend to stress the difference between the approaches. While acknowledging that their philosophies and methodologies are not compatible, I argue that they share a commitment to place 'man' at the centre of 'his' world. More than this, behavioural, humanistic and radical geographers seem to have drawn the same conclusion about that place. Whatever way they choose to express it, there is apparently a 'dialectic' between the individual and the external world.

The rest of this book is an attempt to specify, and tease out, what this 'dialectic' might be like – from a psychoanalytic and geographical perspective. Part II discusses ways in which Freud and Lacan understood the relationship not just between the individual, their internal world and their external world, but also between meaning, identity and power. In the conclusion to Part II, I stage a dialogue between Freud, Lacan and Lefebvre to articulate a framework through which it is possible to reconceptualise the relationship between the subject, society and space. Two sites are identified in this analysis which are seen as having particular significance, theoretically, empirically and politically: the body and the city. Part III illustrates some of the ways in which it might be possible to 'psychoanalyse' space or interpret the psychodynamics of place. It seems to me that this exercise will have no pay off unless the implications for critical theory are also addressed. So, in conclusion, I look to the writings of Fredric Jameson and Frantz Fanon to sketch out the politics of placing the subject.

Part I

GEOGRAPHIES OF THE SUBJECT

2

ENVIRONMENT, BEHAVIOUR, MIND

In as much as the origins of geography, like so many disciplines, reach back to the first vague questioning of man concerning the world around him, most geographical problems have an environmental content.

(Kirk, 1963: 364)

INTRODUCTION

Geographers have always been interested in the 'vague questioning of man concerning the world around him', but they have not always thought of Man in the same way. Geographers have been all too ready to experiment with new ideas and approaches in the search for a better understanding of the relationship between the individual, spatial behaviour and the environment. Commonly, they have ransacked other disciplines, from anthropology to architecture, from psychology to sociology. An interest in overt spatial behaviour led some geographers to consider the ways in which people understood – or perceived – their environments. They argued that man's vague questioning brought about specific answers, and it was on this specific understanding of the world around them that men acted. The place to look for an explanation of spatial behaviour, following this line of argument, was not in the physicality of geography but in the mental pictures that men developed of the world around them. Very quickly, an identifiable and distinct sub-discipline developed in geography, called behavioural geography.

Before moving on, I should say something about the gendering of this language: even today, it is not unusual to find geographers making sweeping comments about Man, but it is remarkably unusual to find discussions of anything but Man or men's interests up to the 1980s. I have left their use of Man intact so that the gendering of their arguments remains explicit and, sometimes, so that I can smuggle in the odd joke. I believe that behavioural geography is constituted through an implicit, undisclosed, oppressive and obstructive masculinism.[1] I will return to this point in the conclusion. It should be further noted that behavioural geography produced a very cosy

1 Gillian Rose does not discuss behavioural geography, but her arguments would be applicable here (G. Rose, 1993).

understanding of the world: there are no social conflicts, no power relationships and certainly nothing shocking. In contrast, psychoanalysis would, I believe, help to uncover relationships of power, to produce a more sensitive and situated knowledge of the world around.

While the tone of this chapter is somewhat critical, it is nevertheless directed towards establishing some kind of common ground between behavioural geography and what might be called a 'psychoanalysis of space'. The suggestion that there is any kind of common ground between psychoanalysis and behavioural geography might seem bizarre or even ludicrous. On the face of it, behavioural geographers have had nothing to do with psychoanalysis and they cannot, therefore, share any of the same concerns. Moreover, it is true that psychoanalysis has been (deliberately or not) avoided or rejected as unsuitable because it does not appear to offer a way to understand overt spatial behaviour. Nevertheless, I believe that behavioural geographers have continually stumbled across psychoanalytic concerns: from their discussions of emotions and motivations, through the analysis of images and the limits to rationality, to their repeated appeals to the 'black box' of the mind and the subconscious. The most important shared concern is with the psychology of the individual and its relationship to both an 'internal' or 'inner' world and an 'external' or 'outer' world. Terms like 'inner' and 'outer' have been put in inverted commas because formulations of behavioural geography and psychoanalysis agree that the 'inner' and 'outer' are in some form of dynamic relationship with one another – although they disagree over how this is to be understood and conceptualised. Indeed, this relationship between an inner world and the outer world is an appropriate place to begin a discussion of behavioural geography's accounts of the relationship between the environment, spatial behaviour and the mind.

PERCEPTION AND THE PHENOMENAL ENVIRONMENT

The work of Kirk provides a clear idea of how geographers began to conceptualise the relationship between the inner and outer worlds. His starting point was the universal need for geographical knowledge. For Kirk (1963), geographers and men have the same basic concern: they want to know where they are, what the world around them is like and how to get around this world.[2] Geography's interest in Man–Environment relationships is the modern expression of this primal wondering, but the environment is now known to be created by human beings themselves, at least in part. Kirk had raised this issue in an earlier paper on the function and methodology of

2 These grand statements about Man's innate needs were common in the 1960s. They were often based on the findings of either anthropology or psychology, which both purported to elicit the universal truths of Man's nature. The terms of reference of these debates were highly circumscribed around the experiences and interests of men (and rarely women). This is a profoundly masculinist and heterosexist imagination (see also G. Rose, 1993).

geography (1952). In this influential paper, he first began to explore the Man–Environment relationship from the perspective of gestalt psychology, although he also drew inspiration from Vidal de la Blache. He argued that the point of historical geography was 'to grasp and reveal at each instant of their duration the complex relations of man, the actors in, and the creators of history with organic and inorganic nature and with the many factors in their physical and biological environment' (1952: 155).

This method, which he believed was sensitive to the complex relations between 'man' and environment, has four constitutive elements: the 'physical environment as a factor in History' (page 155); 'the evolution of landscape' (page 156); 'man as an agent of environmental change' (page 156); and 'the restoration of past geographies' (page 157). However, the evaluation of these alone could not, in Kirk's opinion, provide a comprehensive historical geography, because there is also 'the need to seek values as well as facts, to function as a bridge of ideas, to develop a four-dimensional study, to concern ourselves with action and process as well as distribution, and to restate the problem of environmentalism in some form' (1952: 158).

This search required a new theoretical schema, one which could take into account the subjective experiences of people, while at the same time placing people inside a physical, geographical and social environment, which people nevertheless can alter through their actions. For inspiration, Kirk turned to gestalt psychology, drawing on the work of Kurt Koffka (1929) and Wolfgang Köhler (1929). For Kirk, gestalt psychology was particularly useful because it provided working hypotheses which linked humanity and nature, by concentrating on the ways in which people both perceive the world and order their perceptions into a mental pattern, or 'gestalt'. There is a connection between these patterns and geography:

> A 'Gestalt' is almost precisely what we imply when we define a region in its dynamic aspect, a whole which is something more than the sum of the parts; the psychology of Mackinder, his ability to see things in functioning wholes, to alternate figure and ground when examining a map, and his sense of 'requiredness' are essentially the attributes of the Gestalt Psychologist; but here we are concerned especially with what is termed the psycho-physical field.
>
> (Kirk, 1952: 158)

For Kirk, this psycho-physical field is critical because it connects the internal mental world of people with the external physical world of society and nature: it directly – and in an unmediated way – links people to their environment. From this perspective, the art of historical geography is to track the inner/outer world through time, where this would illuminate the relationship between Man and Nature at any given point in time. Kirk suggests that the psycho-physical field may be understood as the behavioural environment and that this environment closes the gap between Mind and Nature (1952: 159). At the back of this lies a particular relationship between Man and Nature, and Kirk puts this in unambiguously gendered and heterosexual terms: 'the story of man may yet be written not as a conquest but as a marriage' (1952: 160). In his paper on 'Problems of geography' (1963), Kirk refines the story of this relationship between Man and Nature: the marriage between geography and patterns of behaviour involves a contract between the 'inner' Behavioural world and the 'outer' Phenomenal world.

Kirk reiterates that the Man–Nature relationship can be separated into two aspects: the phenomenal and the behavioural. The phenomenal environment is the 'real' world; nevertheless it is also the product of, and the condition of, human action. Thus, the phenomenal environment is the object of perception, but what we choose to see is modified in the behavioural environment: 'The Phenomenal Environment will enter the Behavioural Environment of man, but only in so far as they are perceived by human beings with motives, preferences, modes of thinking, and traditions drawn from their social, cultural context' (1963: 366). Accordingly, the same data may be arranged into different patterns by different people in different places at different times: 'The Behavioural Environment is thus a psycho-physical field in which phenomenal facts are arranged into patterns or structures (gestalten) and acquire values in cultural contexts' (1963: 366).

The Behavioural Environment, because it is shared in communication, is a rational zone in which decisions are converted into actions, but this is filtered by the selective values of culture: 'Facts which exist in the Phenomenal Environment but do not enter the Behavioural Environment of a society have no relevance to rational, spatial behaviour and consequently do not enter into problems of the Geographical Environment' (1963: 367).

Kirk then adds a historical dimension to his argument, asserting that cultures that fail to reproduce and renew themselves fail because of their inability to modify or adapt their world-views – a kind of psychological Darwinism.[3] The important question for Kirk is how to link the Phenomenal and Behavioural Environments in an analysis of the Geographical Environment. At the heart of this problem appeared to be the relationship between the outside world and the inner mind. It was argued that the inner mind perceived the image of the outer world: that is, that 'the image' of the world stood between the inner and the outer worlds. Since 'the image' was the key constituent not only of the psycho-physical field but also of overt spatial behaviour, it ought to become the focus of geographical analysis.

In sum, Kirk provided a theory of the relationship between behaviour and mind that offered an alternative mode of explanation for geographers who were sick of the over-simplified and reductionist models that geography was commonly utilising. Behavioural geographers, as they were to become known, leapt on Kirk's gestalt psychology with some enthusiasm because it provided a human intermediation between overt behavioural responses and

3 Although Kirk cited evidence from ancient civilisations, such rhetoric could not have been innocent of Britain's post-war experience of imperial decline.

the stimulus-providing outside world. The site of intermediation was understood to be perception, where 'perception proceeded according to innate abilities which organised environmental stimuli into coherently organised forms or patterns (*Gestalten*)' (Gold, 1980: 11). Psychology mediated spatial behaviour because it was based on the perception of phenomenal facts: thus, important geographical questions were also questions of psychology.[4]

Perhaps surprisingly,[5] the mode of thought was derived primarily from phenomenology: the primary aim was to gain a sympathetic understanding through describing the whole phenomenon. Nevertheless gestalt psychology came with its own slogans. One of the most important was that 'the whole was greater than the sum of its parts'. This meant that the researcher needed to be aware of the totality of experiences and stimuli acting on an individual in order to make sense of the patterns of their behaviour. Another important distinction was between the perceived environment – in which actual behaviour was rooted – and the physical environment. From this perspective, space has a dual character, as an 'objective environment' and as a 'subjective environment'. It is important to note that the perceived, or behavioural, environment is learned – and this valorised, and continues to valorise, the learning experiences and spatial awareness of children.

Even so, neither gestalt psychology nor phenomenology were to achieve dominance in behavioural geography, which increasingly stressed the analysis of people's images of space. As behavioural geographers turned more and more to the discipline of psychology, they increasingly began to use its dominant mode of scientific knowledge – though their work is still clearly marked by the importance of spatial perception and behaviour. 'The Image' became a central idea, because it was this that was assumed to be the real basis on which people behaved in the real world.

BEHAVIOURAL GEOGRAPHY AND THE IMAGE: MAPS OF THE MIND

From the early 1960s, behavioural geography took off in a big way; partly inside and partly alongside the contemporaneous quantitative revolution. In many ways, behavioural geography was a reaction against the then dominant

4 A parallel, but not identical, case can be made for psychoanalysis.

5 Geography is usually presented as if it was characterised by three discrete philosophical positions (see Gregory, 1978; Johnston, 1986, 1987). This three-fold division, I believe, is grounded in Habermas's identification of three basic forms of knowledge: empirical-analytic, historical-hermeneutic and critical dialectical (1972). This has been translated into geography as the distinction between logical positivisms, humanisms and (western) Marxism. Nevertheless, I suggest that the histories of logical positivism, humanism and Marxism in the production of behavioural geographies are intertwined in their inception, introduction to and development within geography. It was only later that clear divisions developed in geography; even so these knowledges still bear the impress of each other.

normative and mechanistic models of man.[6] At least initially behavioural geographers were surprisingly eclectic, drawing on geographers like Lowenthal, structuralists like Piaget and Lévi-Strauss, cognitive psychologists like Bruner and Werner, and philosophers like Berkeley and Kant. Indeed, Brookfield (1969) and Mercer (1972) saw behavioural geography as a new kind of humanism, because it put a thinking man at the centre of geographical analysis.[7]

Conceptually, mental processes and cognitive representations became of central importance in the work of behavioural geographers because they represented the filter between the mind and behaviour. Significantly, for geography (the discipline), they also implied an interest in the processes leading to observable spatial patterns, but looking at them from the perspective of the people who made them and making theoretically sound statements about those processes (Golledge, Brown and Williamson, 1972: 59). Two key issues, however, came to dominate behavioural geography: first, overt spatial behaviour and, second, in-the-mind environmental perception. For behavioural geographers, understanding spatial patterns and processes was impossible without understanding how the individual 'imaged' the world and made decisions based on that image. This conclusion, it seems to me, is irrefutable, but they were working with a very limited notion of the image and its relationship, not just to psychology, but also to space.[8]

The 'Image' quickly became a pivotal term. Its meaning was derived from the work of Boulding, who 'suggested that over time individuals develop mental impressions of the world (images) through their everyday contacts with the environment and that these images act as the basis for behaviour' (Gold, 1980: 38).

For Boulding, the image was not a fixed picture of the world, but consisted of the growth of any individual's subjective knowledge about the world, whereas behaviour was based on only a part of that knowledge (Boulding, 1956: 5–6). The image therefore was about learning as a cognitive process, but learning in a particular context. First, Boulding placed the image within the valuations which society placed over that cognitive template; and, second, he saw it as a basic means of survival – we learn to adapt or die.[9]

In an analysis of the routine ways in which everyday language is spatially marked by certain metaphors, for him, the image had inseparable personal, emotional and geographical components:

> We think of personal relationships in such terms as near and far, obscure and clear, devious and direct, exalted and commonplace, all of

---

6 See Golledge, 1981: 1,328 and Walmsley and Lewis, 1984: 1.

7 Putting Man at the centre of geographical analysis was, of course, a supposedly defining concern of humanistic geography (see Chapter 3).

8 An alternative, psychoanalytic perspective is presented in Part II and Chapter 7.

9 This argument is reminiscent of Kirk's, although he does not cite Boulding, instead drawing on gestalt psychology, Piaget and Inhelder.

which have strong spatial connotations and are in fact spatial metaphors. It is a challenge indeed to try to think of a metaphor that is not in some sense spatial.

(Boulding, 1973: viii)

Indeed, Boulding went on to suggest that it is almost impossible to communicate without using spatial referents and went on to link these – as Lowenthal (1961) had done – to the limits of the body senses, though these were ranked in order of the spatial information they provided: vision, hearing, touch, smell and taste (with the least spatial information).[10] The image interestingly was not literally a picture in the mind, but a cognitive structure, where a mental process intervened in the relationship between the body and the external physical and social world. And this cognitive structure was routinely expressed in language. This view was echoed by Kevin Lynch (1960). Kevin Lynch's seminal work on the images that people had of Boston, Jersey City and Los Angeles was also an early source of inspiration for geographers. I will return to this work in detail in Chapter 7 below, but it is important to introduce his sense of mental processes now. For him, cognition also was primarily about learning and communicating ideas about the world and was therefore a social activity:

Cognition is an individual process, but its concepts are social creations. We learn to see as we communicate with other people. The most interesting unit of study for environmental cognition may therefore be small, intimate, social groups who are learning to see together, exchanging their feelings, values, categories, memories, hopes, and observations, as they go about their everyday affairs.

(Lynch, 1976: v)

These insights were foundational in the project of behavioural geography, but behavioural geography drew on these ideas in a very particular kind of way. There were basically two strands of thought. The first saw the image as a cognitive structure consisting of limited perceptions of the world which could be elicited, specified and measured; while the second had a much more visual notion of the image suggesting that we carry around in our heads mental maps of the world – which can be elicited, drawn and measured.[11]

---

10 There are two side issues that I would like to raise. First, Boulding here appears to be an early marker of the importance of the visual in contemporary culture. Importantly this may well also help us understand why it is that 'the spatial' has become so important in discussions by cultural critics of the contemporary (see Keith and Pile, 1993a; see also below). Second, Boulding may also be wrong in his ranking of the spatial information derived from the senses, instead it may well be that these senses provide different kinds of spatially marked information. Even – or especially – with closed eyes, the taste of exotic fruit suggests something from elsewhere, whereas the 'taste' of a hamburger might connote the US and roast beef, England.

11 Although behavioural geographers (but especially those working on mental maps) drew heavily on Lynch's *The Image of the City*, he only used the term mental map once (page 88), he never referred to cognitive structures, and he used a qualitative methodology.

I will deal with these approaches in two separate sections, concluding with a section on the relationship between environment, image and behaviour.

The image and bounded rationality

The foundational studies of behavioural geography were concerned to understand people's decision-making.[12] In this first strand of thought, the decision was based on a cognitive structure which was defined by a bounded, or contextual, image of the world. It is here that the problem of the 'boundedness' of everyday life is rendered explicit, and this question has continued to trouble geographers (as we shall see). Thus, White's work evaluated the ways in which flood-plain dwellers assessed the risk of flooding. What emerged from this study was a sense that people were rational, but that this rationality was bounded. This idea was given a formal theoretical status by Wolpert. In his study of the behaviour of Swedish farmers, he argued that they did not make 'optimal' decisions (as they should have done if they really were 'rational economic men') but were 'satisficers': that is, they were much more concerned with avoiding risks than maximising returns. People, moreover, were found to have limited personal abilities. In these ways, then, there were boundaries around farmers' capacity to make rational decisions. Therefore, 'aspiration levels tend to adjust to the attainable, to past achievement levels, and to levels achieved by other individuals with whom he compares himself' (Wolpert, 1964: 545). The farmer could only know and do what the farmer could know and do, and could only be as rational as these limits allowed. Wolpert therefore concerned himself with these limits, which he suggested had five sources: access to resources, the criteria of success, goal orientation, knowledge and the uncertainty of the farm situation.

Meanwhile, Hägerstrand was less concerned with the limits of human rationality than with the limits of the models of spatial behaviour (1965, 1970). He was arguing that behavioural variables needed to be incorporated into models of spatial patterns. While this work was used to suggest the importance of behaviour in the production of spatial patterns, behavioural geography came to define itself in terms of its interest in the individual (see both Gold 1980, and Walmsley and Lewis, 1984). Strictly, behavioural geographers were interested in aggregating individual cognitive processes and overt behaviours into spatial patterns (see, for example, Huff, 1960). This work tended 'to work "backwards" from observable behaviour to make inferences about the mental processes that operate in the mind' (Walmsley and Lewis, 1984: 5). Moreover, it assumed that 'individuals derived information about the environment through perception and evaluated this information in terms of a value system to arrive at a cognitive image in respect of

12 These studies include work by White (1945, 1964), Wolpert (1964, 1965) and Hägerstrand (1965, 1970).

which decisions were made' (Walmsley and Lewis, 1984: 5). While the concept of 'bounded rationality' had some purchase, this strand of thought faltered where it tried to interrelate people's preferences, mental processes and spatial patterns.

The problem seemed to involve the ways people's preferences were organised into spatial images. Attempts were therefore made to render explicit the kinds of images that people held in their heads. This work generated a voluminous literature on what were called mental maps (see Gould, 1973; Gould and White, 1974). Such maps were elicited using techniques known as cognitive mapping (Downs and Stea, 1973a, 1977). It is this work which is characteristic of the second attitude to the image.

The image and mental maps

While many writers used notions of bounded rationality, others preferred to talk of images in terms of cognitive mapping and mental maps. For them, the most useful 'process' for understanding and describing the handling of information in mental processes was cognitive mapping.[13] 'Cognitive mapping is a process composed of a series of physiological transformations by which an individual acquires, codes, stores and recalls, and decodes information about the relative locations and attributes of phenomena in his everyday spatial environment' (Downs and Stea, 1973a: 9).

From this perspective, mental processes included the innate capacity to literally map the world around. Accordingly, this map enabled people to assemble, understand, remember and use spatial environmental information. Even so, 'the actual process of cognitive mapping is very much an unknown quantity' (Walmsley and Lewis, 1984: 8). Behavioural geographers ignored this problem, instead relying on the assumption that 'cognitive maps are convenient sets of shorthand symbols that we all subscribe to, recognise, and employ: these symbols vary from group to group, and individual to individual, resulting from our biases, prejudices, and personal experiences' (Downs and Stea, 1973a: 9).

Cognitive maps and mapping were assumed to be a central component of actual behaviour: people after all need to know where the things they need are, what the area is like and how to get there. This of course is learned – and indeed different practices of cartography have changed the way people think through space (see Massey, 1995). The image is however not the same for everybody, nor does it mean the same thing, nor does it necessarily provide security. Even so, behavioural geographers began to see cognitive mapping as more than simply one (metaphorical) way of organising spatial information: it came to stand for the mechanics of the mind. Mental processes became

---

13 I will follow up the political implications of these issues in Chapter 8 below, through a discussion of Fredric Jameson's appeal to an aesthetic of cognitive mapping (see Jameson, 1984, 1991).

mental mapping, just as spatial images caused spatial behaviour.

Like mapping, cognition was held to have basic spatial characteristics, such as 'a view from above', changes of scale, and symbolic representation (see Downs and Stea, 1973a: 11–12). Like maps, moreover, cognition is incomplete, distorted and schematised, enhanced in some way (augmented, for example, with fictional or mythic beasts), and exhibits differences between groups and between individuals, even within the same group. Maps and cognition betray an instrumental rationality: 'we see the world in the way that we do because it pays us to see it in that way' (Downs and Stea, 1973a: 22).

This identification between the (rational) mind and the (rational) map leaves open a series of questions: most importantly, merely stating that behaviour is based on the image (whether in map form or not; whether cognition is the same as mapping or not) does not solve the relationship between environment, image and behaviour, because it cannot say how or why this particular relationship between environment, image and behaviour exists. Arguments about the ways mental maps are incomplete, distorted and schematised are significant and should be remembered, but they nevertheless fail to grasp people's emotional dynamism, in part seen in the constitutive ambivalence of many, if not all, spatial practices.

Environment, image, behaviour

By the 1970s, a large literature had been developed by behavioural geographers, though their impact was mainly to be seen in the areas of decision-making and choice behaviour, information flows, models of searching and learning, voting behaviour and perception research mainly on hazards, images and mental maps.

The original dichotomy between the phenomenal and behavioural environments had come under close scrutiny and had been found to be wanting. Two environments, for example, became four: Sonnenfeld (1972) suggested that the everyday world should be divided into (1) the geographical – i.e. the world, (2) the operational – i.e. the world encountered by man, (3) the perceptual – i.e. the world aware to man, and (4) the behavioural – i.e. the world eliciting behaviour. Or three: Porteus (1977) divided the world into (1) the phenomenal environment (of physical objects), (2) the personal environment (of the perceived images of the phenomenal environment), and, (3) the contextual environment (of cultural beliefs and expectations).

In these schemas, perception still played the central, articulating role; it was 'an all-pervasive and supremely important intervening variable in any analysis of the spatial expression of human activities' (Golledge, Brown and Williamson, 1972: 73). Though two basic approaches toward the notion of perception developed – bounded rationality and cognitive maps – these shared a view that something mental intervened between behaviour and environment: this was the image.

Both schema also shared another characteristic; while the conscious mind was held to be rational, there was assumed to be an underlying layer – often called the subconscious – which held either irrational or 'unconscious' or biological impulses. This level was unfortunately never theorised, partly because it was assumed to be universal and unalterable and could therefore be assumed away. I say unfortunately partly because this level was obviously a determinant of behaviour and partly because it is one way in which psychoanalysis could have contributed to interpretations not only of overt spatial behaviour, but also of never mentioned covert spatial behaviours.

Instead of following earlier leads from gestalt psychology, phenomenology, Piaget (and psychoanalysis), geographers looked to the supposed explanatory power of cognitive psychology: 'Psychology brought a sense of cognitive process, geography a sense of environment, and the concern switched from an emphasis on static visual images and maps to an exploration of the process of learning and manipulating information about the spatial environment' (Downs and Meyer, 1978: 73).

There were clear overlaps in interests between geographers and psychologists. A brief tour of the journal *Cognitive Psychology* will uncover many articles by psychologists on spatial awareness and behaviour, and these proved very seductive to geographers. While behavioural geographers retained a sense of the importance of spatial images, they broke them down into the reference points of psychology: perception and cognition. It was argued that through the mental faculties of perception and cognition, the individual sensed, felt, perceived, thought about, interpreted, made sense of, made decisions and did; they incorporated motivation, emotion and attitudes into an understanding of overt spatial behaviour, but these referents still excluded that level of the mind which had been called the subconscious (or, in psychoanalytic terms, the unconscious).

By the late 1970s, using cognitive psychology and its own findings, behavioural geography had built up a repertoire of concepts which constituted its working model of the mind. The perceiver was assumed to be tuned into the environment somehow. An individual only comes into contact with stimuli that are perceived, and this is less than the total number of stimuli available in an environment such as a shopping mall. These stimuli are then organised through cognition – which is in part related to the senses of the body, in part to previous experiences of the body and in part to cultural experiences. The underlying principles of spatial behaviour, established in how information is processed and what information people choose to process (Golledge and Rushton, 1976), were found to be influenced by three principles: motivation, emotion and attitudes. Before describing these it is important to note, for the thesis I am presenting, that in their work there was a continual appeal to a level of mental functioning which behavioural geography was unwilling or unable to conceptualise. This level is, however, the very ground on which psychoanalysis operates: underlying motivations, emotions and attitudes are *terrae incognitae* where behavioural geography was not prepared to find itself.

In behavioural geography motivation is 'the force that leads men to seek certain goals in relation to their needs' (Gold, 1980: 21). Gold argued that there were two basic theories of human needs: first was the need to reduce tension; and second was the need to increase tension. These can be seen respectively as pain-minimising and pleasure-maximising; what is remarkable, to me, is how close these come to Freud's suggestion of reality and pleasure principles. Whereas psychoanalysis reveals that these principles are dynamic and mutually-constitutive, behavioural geography tended to see pain-minimising and pleasure-maximising as static (because they are natural survival instincts) and mutually exclusive. They went on to posit a hierarchy of needs, survival needs being more fundamental than social or personal needs (following Maslow, 1954).[14] The question still remained as to whether particular motivations – such as the desire to defend territory – were natural or the result of the social environment.

For the behavioural geographers, emotion was linked to motivation – as in the desire to defend territory (as Freud noted! See Chapter 4). Emotion also incorporated a wide range of states, but 'some emotions accompany motivation ... Other emotions are the result of motivated behaviour' (Gold, 1980: 22). But behavioural geography was not in a position to provide a theory of the gap between motivation and emotion or of the relationships between the two, or even of the contents of motivation and emotion. Alternatively, psychoanalysis could; but really behavioural geographers had become much more interested in the third factor supposedly underlying spatial behaviour, namely attitudes.

'Attitude' became an invaluable concept not only because it was 'the summary of past experiences and perceptions' (Pocock, 1973: 253); but also because it encompassed 'both internal mental life (including cognition, motivation and emotion) and overt behavioural responses' (Gold, 1980: 23); and also because attitudes showed how individuals were confined within the images that a culture permitted them to see, believe in, act on and feel for (Boulding, 1956: 14). Attitudes (like the image) became another catch-all 'location' which stood at the intersection between mind and behaviour, and – because of this – could be used to explain spatial patterns.

From this perspective, attitudes were a predisposition to respond to a particular set of stimuli in a certain way, and when confronted with an unfamiliar situation the individual's attitudes would give him or her a particular set of pre-existing resources to inform their behavioural responses. Attitudes were not simply innate – burned into the circuitry of the brain – they were learned. Nevertheless attitudes were assumed to be relatively constant over time (learning has its limits) because attitudes were felt strongly. And because they were felt strongly, they motivated behaviour, and therefore could be used to explain behaviour: for example, deeply-felt attitudes such as racism or religious bigotry would explain Ku Klux Klan attacks in America's deep south or the troubles in Northern Ireland.

---

14 It is useful to note that humanistic geographers also drew on Maslow's social psychology.

Attitude (like mental maps before) became a metaphor for the whole of mental functioning; it involved the processes of cognition, emotions and the predisposition to behave in a particular way; it included a host of elements such as 'belief, bias, faith, ideology, judgement, opinion, stereotype, and value' (Gold, 1980: 24); it included both mental and behavioural aspects. Nevertheless, this understanding of the relationship between environment, image and behaviour was coming under increasing scrutiny by the late 1970s; in particular, scepticism was being expressed towards the preoccupation with the aggregation of individual's overt spatial behaviour and with the theoretical transparency of middle terms, such as 'the image', 'mental maps' and 'attitude'. Some discussion of these criticisms helps to reveal how far behavioural geography had assumed away 'covert' mental processes, consigning them to an unintelligible zone: the so-called black box.

EARLY CRITIQUES OF BEHAVIOURAL GEOGRAPHY

Burton (1963: 157) and Brookfield (1969: 51) had both predicted that the new perception studies would come to rival the then dominant and self-confident quantitative revolution (see Billinge, Gregory and Martin, 1984). But, by the early 1980s, a range of criticism had already been voiced. From a radical perspective, Massey (1975) argued that because behavioural geographers treated the external world as if it was immutable and given, they reinforced the status quo: 'The reason that an empiricist approach to behaviour is incorrect is that it petrifies existing "facts" into external laws. Forms of behaviour can never be taken, theoretically, as given; they are always "produced" – that is they are an outcome of the structured context in which they occur' (1975: 202).

Similarly Cox (1981) claimed that, in relation to the mechanistic models that dominated logical positivism and systems theory, behavioural research (despite its origins) was simply 'business as usual' and in no way constituted a radical break with these approaches. Cox argued that behavioural geography had not managed to escape the assumptions of bourgeois, classical economics: i.e. rational economic man (again, despite its claims). These criticisms seem well-founded, given statements like this: 'spatial behaviour, exactly as any other behaviour, is determined by preferences only' and preferences 'are all we need to predict spatial behaviour in a system' (Rushton, 1969: 400).

Doubts were also expressed about behavioural geography's ability to theorise the relationship between overt spatial behaviour and the processes that caused that behaviour (see Olsson, 1969). Further, Thrift (1981) thought that the environment was dealt with one-dimensionally: the whole environment was dealt with through the flat concept of perception, which could not account for deep social structures. Similarly, Gold and Goody (1984) pointed out that by focusing empirical and theoretical work on the psychology of the individual, behavioural geography had therefore failed to do justice to social behaviour – and therefore could not explain overt spatial patterns.

Ironically, the most telling criticisms were provided by a supposedly sympathetic critique. It was Bunting and Guelke who best caught the mood of scepticism surrounding behavioural geography (Bunting and Guelke, 1979). They criticised behavioural geography's concentration on 'The Image' and this was blamed for the theoretical weakness of behavioural geography. They claimed that behavioural geography assumed that people behaved according to the environment as it is perceived and interpreted; it was therefore focused on the individual's revealed views, as opposed to those of a group or organisation. There were two central assumptions that behavioural geographers never challenged: first, that images could be measured; and, second, that there was a strong link between revealed perceptions and actual real-world behaviour: 'Behavioural and perception geographers have assumed that environmental images can be measured accurately and that there are strong relationships between environmental images and actual behaviour' (Bunting and Guelke, 1979: 448).

Bunting and Guelke were not trying to undermine behavioural geography but to suggest that it is better to avoid these assumptions by looking at observable spatial behaviour and then to infer what is going on in the mind, rather than to postulate mental process and then look for the processes and their resultant phenomena. They were particularly scathing about the promise provided by psychological models and statistical techniques for an understanding of overt spatial behaviour. Like the radical critique, they also noted that behavioural geography had failed to understand that 'man's environmental behaviour is not one-dimensional: it is multifaceted. It involves economic, social and political considerations as well as environmental ones' (1979: 461). Nevertheless, behavioural geographers were quite sympathetic to this sympathetic critique: three replies quickly followed (by Rushton, 1979, Saarinen, 1979 and Downs, 1979).

Basically, these replies asserted that Bunting and Guelke were attacking a caricature of behavioural geography, but that they had a point. Rushton, for example, agreed that behavioural geographers had assumed that there was a strong relationship between perception and behaviour, but that what was required was greater consideration of the decision-making process. While Rushton refused to accept the reduction of behavioural geography to perception studies, Saarinen refused to accept Bunting and Guelke's conflation of perception studies and mental maps. Downs was more ambivalent: 'that the patient is ailing is beyond question; that the patient is failing is debatable' (1979: 468). These muted responses seemed to share Rushton's disquiet that behavioural geography's measurement of perception, images and attitudes had been 'to no apparent purpose' (1979: 464).

A more assertive riposte was written by Golledge (1981). He argued that, in critiques of behavioural geography, there were several misconceptions, misinterpretations and misrepresentations. Golledge identified nine myths; some of which, he admitted, were actual tendencies within behavioural geography which were in danger of becoming a reality. These myths were that behavioural geography:

1 'can deal with large population aggregates (such as a census) without losing its essence';
2 'is just as valid for studying the activities of inanimate systems as it is for sensate behaviour';
3 needs 'to focus entirely on overt facts – for these represent the "true" preferences and choices of decision-makers';
4 'is "behaviouristic"' or 'implies a phenomenological or idealist viewpoint';
5 has failed 'to critically examine the assumption that environmental images exist in people's minds';
6 has provided 'no evidence to show relations between perceptions and behaviour';
7 its 'research on images, perceptions and preferences has contributed nothing to the understanding of actual behaviour patterns';
8 has provided 'no attempt to cross-validate results obtained'; and,
9 its 'commitment to statistical testing and interdisciplinary approaches gives the approach its distinctiveness rather than a recognition of the "internal" bases of human spatial behaviour'.

(Golledge, 1981: 1325–1326)

To refute these myths/tendencies, Golledge began by providing a potted history of behavioural geography. This is important because he was reminding those inside and outside behavioural geography that it was primarily concerned with providing an explanatory layer grounded in the everyday experiences of sensing, thinking and behaving people. It was intended that variables derived from behavioural geography should supplement those derived from other geographers' explanatory schema. The goal of behavioural geography was to provide a new perspective which stressed, in association with psychology, human variables such as information, risk assessment, environmental images, attitudes and revealed preferences. Such variables were to be understood in terms of mental processes such as learning, perception, cognition, attitude formation, and so on: 'the approach emphasises human actors in the complex interacting system of human and physical systems' (1981: 1328). So behavioural geography was never meant to be a stand-alone approach to geographical problems and, therefore, could not be solely responsible for its own weaknesses.

For Golledge, behavioural geography's strength was that it could explain spatial phenomena by uncovering the reasons why people behaved in certain ways as opposed to merely describing the overt patterns of behaviour. Because behavioural geography was concerned with the reasons behind behaviour, it stressed individual differences, seeking to make generalisations not through aggregation, but through tried and tested hypotheses. According to Golledge, then, behavioural geography works from the individual upwards to the social: 'one of the major contributions of behavioural work was, in fact, to point out to geographers that their classical models assumed stereotyped, mechanical, or habitual responses by individuals to spatial and social situations and structures' (Golledge, 1981: 1329).

There is an explicit engagement here with criticisms pitched at a philosophical level. This is more fully spelt out in a later article (Couclelis and Golledge, 1983). Couclelis and Golledge's argument relies on setting positivism within its own limitations. They believed that positivism should be seen as one form of knowledge, with a particular emphasis on openness, clarity, self-criticism and continual questioning. One reason for the acknowledgement of these limitations was the problem of the mind, which could not be scientifically evaluated in the way that, for example, a glacier can: 'What cognitive researchers are searching for is not some a priori given "object" existing in their subjects' minds, but the result of the subtle interaction between particular scientific methodologies and that undefinable something which we assume governs conscious decision making and behavior' (Couclelis and Golledge, 1983: 333).

Even so a scientific goal was to be maintained, though certain mistakes – resulting from the inappropriate use of logical positivist procedures – were to be avoided. Behavioural geography would avoid deterministic models first by not calculating averages for aggregate behaviour, second by not building macro-scale models, third by not using deterministic stimulus-response models, and finally by not concerning itself only with overt behaviour, but also by studying the decision-making processes as well.

In the end, although still optimistic about the future of behavioural geography, Golledge was forced to admit some kind of defeat: 'the "powerful new theory" promised by early behavioural researchers (Cox and Golledge, 1969) has not been articulated as a result of behavioural work' (Golledge, 1981: 1338).

It is worthwhile asking why this is, and why it remains true. For me, these criticisms rightly highlighted deficiencies within behavioural geography and the limits to its understanding of spatial patterns. However, there are parts of the project which I would like to argue for: sadly, these are also positions which were abandoned early on, in the scramble for scientific credibility. I would like to argue for a reconceptualisation of 'The Image', boundaries and rationality, and partly relocating emotions, feelings and ideas. Moreover, putting 'psychology' at the centre of an understanding of the relationship between the individual, the environment and behaviour seems to me to be a reasonable thing to do. Problems remain. If the project was about the relationship between the individual – as mind and body – and the world around, then behavioural geography needed to be aware of its underlying assumptions about the 'black box' of mental processes. For me, psychoanalysis offers the prospect of moving beyond these undisclosed, undynamic 'black boxes', which ultimately could only produce static descriptions of overt spatial behaviour.

THE NATURE OF MAN: OF RATS, ROBOTS AND THE UNKNOWN

Behavioural geographers argued that the quantitative revolution had little appreciation of the mind. In the first instance behavioural geography itself also worked with very rudimentary assumptions about the mind and behaviour. Even Kirk's paper had not put together a theory of the mind in itself: 'The mind was thought of very much as a black box and researchers had only the most primitive notions as to how the perception process worked' (Walmsley and Lewis, 1984: 4).

Nevertheless, behavioural geography was an attempt to shine a light into the 'black' box and develop a 'white' box concept of the mind.[15] The 'white light' was perception (see Downs, 1970: 68). The study of perception illuminated the black box by developing an understanding of the ways in which knowledge was stored and ordered in the mind. Certain basic assumptions about the nature of man began to develop. But there has been a consistent reluctance to think about these 'models of man', even though particular models of man were necessarily adopted (Agnew and Duncan, 1981: 45).

In this interpretation of the mind, two levels were assumed to exist: the conscious and the subconscious, or the image and the schemata (Tuan, 1975). Where the image was associated with creativity and imagination, the schemata was associated with everyday experience. Where the image required effort of memory, the schemata is 'employed almost subconsciously' (Gold, 1980: 42). The image was to be found in consciousness, but the schemata was the supposedly unalterable template which underlay, and determined the parameters of, behaviour. Even so, the relationship between the image and the fixed template was still fluid – because the individual was always encountering new situations and learning from experience.

Despite warnings to the contrary by Burnett (1976) and Gould (1976), this 'two level' conceptualisation of the nature of the mind was smuggled into geography's analysis of man–environment relationships in a variety of guises. For me, there is a two-fold problem: first, that the nature of the lower level (be it schemata, subconscious, template, or whatever) is rarely, if ever, discussed; second, that this lower level is presumed to be universal or unalterable – and therefore not part and parcel of an understanding of differences in spatial behaviour. Psychoanalysis, it seems to me, will provide an alternative version of this 'lower level' and its relationship to other 'levels' of the inner and outer worlds, but first it is necessary to disclose the implicit assumptions about the workings of the mind.

Here, I have identified two fundamental models of man which underlie

15 The colouring of this metaphor can be no accident – white being associated with humanity, the open mind, rationality and self-control, 'black' associated with animalism, the body, irrationality and emotions (see the discussion of Fanon's ideas in Chapter 8 below).

the development of analytic behavioural geography: namely, Ratomorphic and Robomorphic Man. I will deal with each of these in turn.

Ratomorphic Man

This model could have been called many different things – such as biomorphic, genomorphic, even sexomorphic – but the term ratomorphic is intended to highlight behavioural geography's relationship to stimulus-response psychology, which is otherwise known as behaviourism. Behaviourism was established by J. B. Watson and B. F. Skinner and it was concerned with the relationship between a stimulus and its response. It was contemporaneous 'with "environmental determinism" in geography and with the assumptions connected with "mass society" in sociology and "economic man" in classical economics' (Gold, 1980: 9) – all of which were wholeheartedly condemned by geographers.

From the beginning, behavioural geographers saw behaviourism as something to be avoided because they believed that man was thought unable to transcend his reflexes and was therefore at the mercy of external stimuli (Agnew and Duncan, 1981: 51). Men were treated as if they were laboratory rats – and to treat men as if they were at the mercy of their biology and external influences was obviously daft. In this vein, Walmsley and Lewis derided mainstream psychology for dealing with 'ratomorphic man' (Walmsley and Lewis, 1984: 10). Importantly, this caricature of behaviourism proved that behavioural geographers were not guilty of being behaviourists. Instead, it was claimed, behavioural geographers had always used psychological techniques which did not rely on rat-like assumptions: e.g. based on Kelly's personal construct theory.[16] However, this 'autobiography' of behavioural geography underestimates not only the invidious and pervasive way in which the model of ratomorphic man was actually imported into geography but also the complexity of this account of the nature of man.

An important theoretical resource in early behavioural geography was a paper by Tolman, in which he spent some time discussing the nature of cognitive maps in rats (1948). In this paper, Tolman discussed stimulus-response theories of behaviour. His version of these theories is more subtle and provides a less obviously daft theory of human behaviour. Where the caricatures had it that the basic assumption was that people simply respond, without any will, to stimuli; stimulus-response theories conceptualised the ways in which animals *learned* particular behaviour patterns.

Tolman outlined two kinds of stimulus-response theory, whereas geographers even into the 1980s only saw one. Tolman argued that there was a difference between what he derided as the 'telephone switchboard' theory and the 'field theorists' (of which he was one). In the 'switchboard' model, the mind acted as a switchboard operator who keeps a particular line open

16 See Kelly, 1955 and Hudson, 1980.

after a call has been placed. A particular stimulus (incoming call) and its response (placing the line) strengthens a connection in the brain, so that response becomes more and more likely over time, whereas other responses become less and less likely. Different versions of this school disagreed as to why this happened: one suggesting that the connection was kept open because of the frequency of the stimulus and response, the other suggesting that it was because of satisfaction obtained by reacting to a stimulus in a particular way. For Tolman, the switchboard was an inappropriate metaphor for the functioning of the brain. Instead, he preferred to believe that 'in the course of learning something like a field map of the environment gets established in the rat's brain' (Tolman, 1948: 31).

The term 'map', once found by geographers, immediately endeared Tolman to them. Even while most geographers found the mechanistic switchboard-like stimulus-response behaviourism abhorrent, they succeeded in importing this model in the bastard form of cognitive mapping. This may be illustrated by the brief exchange between Getis and Boots (1971) and Stea (1973). Getis and Boots argued that geographers had used psychological theories without realising that their hypotheses about behaviour were based on laboratory observations of two different kinds of rat: Rattus Norvegicus (a particularly aggressive version of the bog standard rat) and white rats (strains of which were specially bred for laboratory use). Getis and Boots doubted that geographers could really benefit from this research and they backed up this assertion by quoting Tolman himself as saying 'all psychology is at bottom cultural: I stand for rat psychology: the rat is not cultural' (Tolman, 1946; cited by Getis and Boots, 1971: 11). That is, Getis and Boots argue, man is cognitive: 'rat behaviour, unlike most of man's behaviour, is unlearned or culture-free' (1971: 13). Men, apparently, are not rats.

Rats therefore cannot be used to formulate hypotheses about man's behaviour, except – and this is important – in situations where culture does not influence man's behaviour. Men do behave like rats – unless trained not to. Stea was moved to respond to this allegation: while men, he confirmed, were not rats, nevertheless hypotheses derived from animal psychology could usefully be reinterpreted. For him, the study of rats had 'serendipitously uncovered some interesting aspects of spatial behaviour' (Stea, 1973: 106) – and, by extension, of the spatial behaviour of man. Indeed, Stea was prepared to go a little further: interestingly, he suggested that 'an optimum environment for rats ... was remarkably similar to that of the garden city concept' (Stea, 1973: 109). The production of urban space could be planned, accordingly, using information gained about rat spatial behaviour. Implicit in this argument is the importance of the cognitive map in intervening in the behaviour of laboratory rats and real men.[17]

---

17 It is probably useful to remind ourselves at this point that studies of the history of science have shown that the social values of scientists are readily transferred into their understandings of other animals, see for example Donna Haraway's essays on primate studies (1989, 1991).

There was here a double misrecognition by geographers: first, as I have said, they failed to see cognitive maps as a part of the stimulus-response model of human behaviour and, thereby, overlooked how rat-like their assumptions about cognitive structures were; and, second, they also missed the social comment in Tolman's article. This last point may seem wide of the mark, but something more was going on in this article than behavioural geographers ever admitted. This story has to wait, however, until the end of this chapter. As I have said, geographers consciously found stimulus-response models abhorrent – they looked for something as scientific, but less obviously daft and less vague. Given their predilection for mathematical models and the frameworks of scientific methodology (see Harvey, 1969a, 1969b), it is not surprising that the real 'hero' of spatial analysis should resemble a robot – neither daft nor vague.

Robomorphic Man

Again, many different epithets spring to mind in considering this model of man: for example, stochastomorphic, mathomorphic, logomorphic. I settled on a term derived from the film, *Robocop*, who was half man, half machine, all cop. Robomorphic man is the carrier of the law – Logos, logic boards and all – a cyborg, neither man nor machine, but both: '*Man is taken to be a complex information-processing system*' (Downs, 1970: 87, emphasis in original).

This more or less sophisticated information-processing system operated according to specific rules. From this perspective, people's behaviour was to be explained through various kinds of normative model – whether they were founded in principles of mathematics or logic or probability or stochastic learning.[18] For example, Pred's (1967) 'behavioural matrix' relied on stochastic assumptions about behaviour. Pred suggested that a graph could be constructed with 'optimal solution' at the top of one axis and 'perfect knowledge' at the top of the other. The first axis would represent an individual's ability to use information, while the other axis would represent the quality and quantity of information. For each decision to locate somewhere, an individual could be placed at a point on the graph. He argued that the chance of making the optimum locational choice would depend on the ability of the decision-maker and the information available.

Similarly, Harvey (1966) argued that behavioural constraints ought to be seen 'as a random disturbance effect upon the ideal patterns generated by classical theory . . . where the random component summarizes the imperfections inherent in the decision-making process' (1966: 82).[19]

It remains in doubt as to whether this is a more realistic model of man than either the rat or rational economic man – either way these are hardly

18 See for example Harvey, 1966, Golledge, 1967 and Rushton, 1969.

19 This is something of a gratuitous quotation, but it is remarkable that in much of Harvey's Marxist theorising, people only turn up as deviations from classical theory. There are, I would suggest, remarkable continuities between Harvey's logical positivist and Marxist models of man.

flattering. Instead of the cognitive process being the intervening variable between the external world and overt spatial behaviour, man was made of imperfect random components: robots never seem to work in the way they should (from *Metropolis* to *Alien*). Robomorphic Man even became the intervening imperfect variable between the ideal world and actual spatial patterns (see Downs, 1970: 68)!

In a later formulation, robomorphic man takes a much more mathematical form. For example, Pocock and Hudson suggested that behaviour is a function of the relationship between the person and the environment and that this can be expressed as a mathematical equation: i.e. $B = f(P, E)$, where (P, E) is the life space or the behavioural setting (from Pocock and Hudson, 1978: 5). Similarly, Couclelis and Golledge pointed out that 'people's values and emotions must be facts for the behavioural geographer, for they causally determine other facts' such as environmental cognition and actual spatial behaviour (1983: 333). 'The picture that emerges is one in which mind and world are in constant dynamic interaction, the fact becoming part of a mental construct that will shape the next fact as both the things in the world and the mind's experience of these things develop' (1983: 333).

Behavioural geography has found, then, that the mind is an unknown and complex entity in dynamic and constant flux within itself and between it and the external world – which is not real but perceived. How is this 'constant dynamic interaction' to be conceived? By turning to the dynamic psychology of Freud – or mathematics? By turning to mathematics. Couclelis and Golledge argue that the idea of a logico-mathematical structure is 'the most powerful template for drawing out the outlines of that part of existence that is accessible to logical thought' (1983: 334). The problem here is that this assumes that the mind has a logical template, and that this is the same in every way as the artificial language of mathematics. Moreover, even if this is right it only gives the researcher access to one part of mental functioning: the logical.

While it is openly acknowledged that the logico-mathematical structure has to exclude 'prelinguistic realities, internal relations present in unanalyzable primitives,[20] global states of consciousness, or events in continual flux' (1983: 335), this remains a problem. Robomorphic man must be able to speak, he must be civilised, he must be an individual and he must be rational and sane. Robomorphic man appears to be white, western and middle class. Couclelis and Golledge's model, they claim, is purely formal, requiring 'no particular semantic commitment about the real' (1983: 335). This is duplicitous. Supposedly neutral science legitimates a corrupt model of man – which defines *who* is on the inside and *who* is to be kept outside, while blissfully ignorant that it does this.

---

20 I do not know what they mean by 'unanalyzable primitives', but it is plausible that this is a racist comment – see the discussion of Freud's 'dark continents' remark in Chapter 6 below. In any case, their Robomorphic Man precludes 'realities' which psychoanalysis includes.

There is no logical reason why the robomorphic model would be any better able to predict overt spatial behaviour than any other model of man, in part, for two reasons: first, like the others, it uses aggregated behavioural data; and, second, like the others, it takes no account of real world circumstances. Moreover, in both these versions of man, ratomorphic and robomorphic, it is consistently assumed that more is going on than the model can account for – there is, inescapably, an unknown quality to the mind. In behavioural geography, there is a recurrent encounter with an area of the mind with which it refuses to deal: that part of existence which is not logical or rational or slow enough to be mapped. This unknown place is most often called the subconscious or a black box. I would call it the unconscious.

The unknown site of the mind

Even as the 1960s gave way to the 1970s, the possibility that behavioural geography would fail to deliver a 'human' theory of the relationship between the mind, environment and spatial behaviour was being noticed and assessed: 'Part of the problem has been our inability (unlikely to be resolved in the near future) to specify precisely what goes on inside the "black box" of the human mind intervening between stimulus and response' (Stea and Downs, 1970: 4).

Behavioural geography's inability to specify the contents of the black box stemmed in part from the problem of how the mind worked. For example, Pocock argued that, when an individual encounters the world, 'only a fraction of the total stimuli is consciously selected and processed' (1973: 252), which begs questions not only about what goes on in the conscious but also about unconscious selection and processing – as he himself argues: 'the input is swelled unconsciously by subliminal perception' (1973: 252). And there the matter is left. Pocock instead suggests that there are four basic components to perception – physiology, psychology, culture and mood – the unconscious has slipped his mind. Meanwhile, the insides of the black box seemed to contain the illogicalities of human existence: that is, those parts of the mind which are impulsive, random or irrational. The black box became a kind of dustbin for all the troublesome elements of being alive which prevented the one-to-one mapping of mental processes onto overt spatial behaviour, elements such as feelings and sensations.

According to Lynch, writing in a foreword to a book by geographers, 'feelings and ideas are not merely troublesome intervening variables that must be passed through in order to understand visible behaviour . . . Feelings and ideas, and the actions and sensations that are a part of them, are what it is like to be alive' (1976: viii).

The concepts of behavioural geography were, however, unable to cope with the complexity and dynamism of 'what it is like to be alive'. While behavioural geographers had noted that the mental mechanics of perception did not explain everything, they had assumed away the deeper patterns which might underlie observable behaviour and recoverable images. Although psychology, physiology, culture and mood were all assumed to influence a person's cognitive map, this was seen as too difficult to test, and therefore prove, scientifically (Pocock, 1973). Tuan put the problem this way: 'though geographers profess an interest in the mind their focus of study remains observable behaviour. The mind appears as a sort of ghost in the machine that makes it easier to understand certain kinds of human action' (Tuan, 1975: 205).[21]

Behavioural studies concentrated on individuals, comparing and contrasting the differences between the mental images and the real world, while positing a mutually interacting relationship between the mental map and the actual map of the lived world. The nature of this mutual interaction was entirely subsumed under concepts such as 'the image' – flat and lifeless, a ghostly surface covering the machine. So, 'a perceived environment can be thought of as a monistic surface on which decisions are based' (Walmsley and Lewis, 1984: 9).

Walmsley and Lewis had actually drawn the idea that the perceived environment was a monistic surface from Brookfield (1969). Closer inspection of this passage reveals a very different sense of the perceived world: 'The perceived environment is complex, monistic, distorted and discontinuous, unstable and full of interwoven irrelevancies; its complexities may in sum be less than that of the real environment, but it is far less easy to separate into discrete parts for analysis' (Brookfield, 1969: 74).

Brookfield starts out, I would say promisingly, by asserting that the perceived environment is full of irrelevancies, complexities and discontinuities, but these difficulties he says have to be put to one side – after all, it is impossible to find out what is really going on inside someone's head (1969: 66). The mind was always just out of reach, a ghostly presence which continued to haunt behavioural geography, but which could not be exorcised. Brookfield's remarks are characteristic of the way behavioural geography attempted to sidestep its own epistemological, ontological and empirical assumptions: how can the perceived world be either more or less complex than the real world, or how can the perceived world be monistic and discontinuous and full of irrelevancies, or who decides what is monistic or distorted or irrelevant and how? For Lynch such questions are thrown up because mental maps are presumed to be a distortion of the real environment. He countered that mental images were real: perhaps a psychoanalytic appreciation of the relationship between the subject, space and the world was not so far away from the project of behavioural geography after all.

While most geographers would surely agree with Brookfield when he argued that 'we cannot pretend to understand man on the earth without some knowledge of what is in the mind of man' (1969: 75), behavioural geography by its own admission failed to provide this understanding. It seems a shame

---

21 Behavioural geographers were aware of the problem: Tuan is quoted by Downs and Meyer, 1978: 69.

that few geographers would suggest that psychoanalysis might help: it is useful to conclude on this point.

CONCLUSION

> The central assumption of behavioural geography is that human beings respond to their environment as it is perceived and interpreted through previous experience and knowledge. *But* interpretation, experience, and even knowledge are clearly functions of social and cultural values and constraints of personal likes and dislikes, of memory, affect, emotion, fears and beliefs, prejudices, misconceptions, mental capacities, habits and expectations, along with all the institutional, economic, and physical factors that characterize the public, 'objective' environment.
>
> (Couclelis and Golledge, 1983: 333, emphasis added)

This is a big 'but': clearly human beings respond to their perceived environment, but as Couclelis and Golledge suggest their interpretations of their surrounding environments (however named) are subject to other 'functions' and 'factors'. The question for behavioural geography was how to take account of these 'functions' and 'factors'. In the main, they either ignored them or treated them as independent variables. So, for example, factors such as class, race and sex were treated as biases on the individual's perceptual environment: they could have learned much from a closer reading of Tolman's arguments about rats' mental maps. While he surmised that rats build up a cognitive map of their environment, Tolman also recognised 'the largely active selective character in the rat's building up of his cognitive map. He often has to look actively for the significant stimuli in order to form his map and does not merely passively receive and react to all the stimuli which are physically present' (1948: 41).

The rat does not simply deploy some mental map out of thin air: the rat actively creates a mental map. Without the distorting influences of social and cultural values, without the added complications of 'personality', without the context of institutional factors, the rat makes its own mind up. The problem here is that the rat's cognitive map – which in behavioural geography functions as the level of understanding and explanation – *is* actively created by some other level: the rat selects, the rat builds up, another part of rat psychology is doing the thinking. Unfortunately behavioural geography did not recognise this other place in the rat's psychology, let alone in Man's. In Tolman's account, rat psychology is split into at least two parts: the part that selects and builds mental maps, and the part that functions through the mental map.

The suspicion that Tolman might be thinking about rat psychology psychoanalytically may be one interpretation too far, but what can be substantiated is that when Tolman boldly extended his analysis to include human behaviour he was forced to draw on another theorist of the mind:

Freud. I will detail Tolman's account of psychoanalysis in the Introduction to Part II, but for now it is useful to note that he was trying to get to grips with certain ways in which stimuli are responded to by human beings. Following Freud, he argued that human beings respond to stimuli according to two related but opposed principles: for Tolman, at least, ambivalence is constitutive of human responses to stimuli, whether their source is their internal or external world. This is the beginnings of a dynamic interpretation not just of human psychology but also of human cognitive mapping.

Had behavioural geography followed this lead, it would have been able to recognise that the 'factors' and 'functions' which Couclelis and Golledge list are not just the contexts within which people perceive and interpret the world, they are also actively created by the perceptions and interpretations of thinking, feeling, acting human beings. External and internal 'functions' and 'factors' do not just impinge on perception, they actively constitute perception and are actively constituted by it. If rat psychology is split, then it seems fair to suggest (without human arrogance) that human psychology is split too. Psychoanalysis presumes that individual psychology is both internally full of life (and death) and in a constantly unfolding relationship with the external world, and neither the internal world nor the external world are presumed to have an essence which is reducible either to biology or to the social.

Part of behavioural geography's failure to recognise the mutual interaction between mind and environment was its insistence on keeping them apart. Concepts such as 'the image' or 'mental maps' functioned to purify human essences (whether located in the subconscious or in the template) of social influences. It acted as a gatekeeper, preventing the switching between the personal and the social, between the irrational and the rational, between the individual and his influences. Behavioural geography consistently operated through what is actually a radically unstable set of dichotomies, such as those between the external and internal worlds, between the public world and the private world, between the subjective (perceptual) and the objective (phenomenal) world, and between Mind and Nature. By separating the analysis of human–environment interactions into two discrete and autonomous entities, behavioural geography foreclosed any sense not only of the relationship between them (closing this account on the standing of transparent middle 'men'), but also that these terms might themselves be signifiers of a psycho-social field. From this perspective, behavioural geography described the categories of perception which it had already presumed existed in advance: it saw only what it expected to see, and did no better than describe it. All of which, it may be said, is hardly subversive.

It is significant that behavioural geographers have been happy to use the term 'Man' without, for example, noting Lynch's discussion of differences between men (1960) or those between men and women, except in derogatory ways. Behavioural geography posited a 'perceptual environment' which stood between the body and the world, it could not see that subjectivity might be forged in the interaction between differences of gender, sexuality, class, race, or that in this maelstrom of personal and social interaction, the subject might have highly complex, fluid and unstable relationships to 'images' or 'mental maps', or that these relationships might vary over any individual's lifespan (i.e. the learning curve isn't always as predictable as it should be, just like robots really).

Through its concepts, behavioural geography quickly converted personal and social interactions into the natural faculties of perception and cognition. Consumer preferences are cosy things with easy regularities: they are not related to the commodification of things, they are not situated within a labour theory of value; consumers are never shocked and they are never shocking; nothing nasty or horrible ever happens and there is no desire. Everything in behavioural geography is on the surface, whereas psychoanalysis is not only able to specify this 'surface', but also to ask questions about what lies beneath (if anything), how these relations of surface and depth, as well as inside and outside, are constituted. From this perspective, consumers might be represented by some surprising figures, like 'the flâneur' or maybe 'the prostitute'.

Psychoanalysis offers the possibility of aiding a project concerned with describing, interpreting and explaining overt spatial behaviour, concerned with putting people at the centre of geographical analysis. But it does not end there. Psychoanalytic notions of the unconscious suggest that the 'behavioural environment' is not just related to overt spatial practices, but to spatial relations in the unconscious; that is, to what might be called covert spatial behaviour. Connections to the world are not just about the arrangement of the psycho-physical field into observable patterns, they are also about other things. It is partly for this reason that geographers dissatisfied with behavioural geography's inability to let go of reductionist and deterministic models of the mind turned towards notions of 'human agency'; unfortunately, these geographers also turned away from discussions of human psychology, and psychoanalysis disappeared over the horizon (for a while).

3

GEOGRAPHIES OF HUMAN AGENCY

INTRODUCTION

Behavioural geographers had instigated a search for models of understanding spatial patterns which did not assume that all people behaved in exactly the same way in any given set of circumstances. Concepts such as the Phenomenal Environment were offered as a way of describing the physical reality of the external world, while it was suggested that the ways in which 'phenomenal' facts entered consciousness and were arranged into patterns and acquired social value could be termed the Behavioural Environment (by Kirk, 1963). The physical reality of the world was only important in so far as it became a part of the psycho-physical field of the Behavioural Environment. This, then, was a people-centred understanding of the relationship between people, the environment and behaviour. Despite these origins and despite their wishes, behavioural geographers developed their interpretations within the framework of logical positivism and scientific methodology.[1] Subterranean understandings of the workings of the mind were smuggled in as geographers looked to science to explain the overt spatial behaviour of individuals and groups. While it was denied that Man was either rational or economic, (implicitly) he was a rat or a robot, though with all the imperfections these 'characters' imply. Ironically, behavioural geography had placed Man at the centre of a world in which he seemed to have no place: in the exploration of *terrae incognitae*, men had somehow lost their hearts and minds.

Geographers interested in 'Man's' vague questioning of the environment around, and in the unknown lands of the hearts and minds of 'men', began to look for other philosophical and methodological perspectives to ground their research.[2] Commonly, circumstances in the United States are also cited as a cause of geographers' dissatisfaction with logical positivism: the late 1960s was marked by the civil rights movement, the women's movement, anti-Vietnam war protests and the power of beautiful flowers. Of course, geographers' dislike of scientific methodologies and the existence of

---

1 Indeed, it was designed to overcome the problems inherent in logical positivism (see Cox and Golledge, 1969).

2 The caveats stated at the beginning to Chapter 2 over the terms 'Man' and 'men' still apply.

alternatives within geography predate the 1960s. Indeed, behavioural geography seems to have spawned its own worst discontents: for example, Yi-Fu Tuan looked to phenomenology, David Harvey to Marx, Gunnar Olsson to linguistic philosophers, and Allan Pred to a developing time geography. Curiously, early articles by these thinkers are still cited by behavioural geographers as if they had written nothing since the mid-1970s; as if the reasons why they had abandoned a full-throated behavioural position were irrelevant.

It is usually suggested that opposition to logical positivist and behavioural geography fell in to two basic camps (each camp occupied by related, but not necessarily friendly, factions): the humanistic and radical approaches.[3] If the differences between humanistic and radical approaches can be summed up in a sentence, then the humanists were interested in philosophies of meaning, while the radical geographers wanted to critique structures of power in society.[4] Sadly, these alternative perspectives on the place of people in the world neglected to discuss 'the mind', perhaps because of their desire to mark out the difference between their geographies and those of the logical positivists. Even so, in this chapter I am interested in the ways in which the philosophies of meaning and the critiques of power understood, albeit implicitly, human psychology: a key co-ordinate in the (contested) map of human subjectivity for both perspectives was 'agency' (see, for example, Duncan and Ley, 1982).[5]

Behavioural geographers, meanwhile, were not isolated from these new approaches to meaning and power, partly as they became increasingly aware of how far their 'new humanism' had come to mirror the 'old science'. As some began to recognise the limits of logical positivist presumptions, they tried to hold the project together by stressing the commonalities between the 'new humanism' and an unfolding humanistic geography. Thus, Couclelis and Golledge claimed that they were interested in both perception, in 'the intimate, ongoing, two-way relationship between person and world' (1983: 336), and in the meanings, feelings and emotions of people. There were other responses. Instead of pushing behavioural geography as a kind of humanism, Walmsley (1974) turned the problem on its head and argued that phenomenology was a form of science.[6] The difference between behavioural geography and phenomenology was, in this light, merely one of emphasis. Similarly,

3 See Gregory, 1978; Johnston, 1986, 1987 and Unwin, 1992.

4 I will not have space and time to develop this complex story, but it should be noted that humanistic geography mainly drew on phenomenology, existentialism and idealism (see Harvey and Holly, 1981), while radical geographers were principally inspired by Marx, Kropotkin and Reclus (see Peet, 1977) – rather than the civil rights and women's movements.

5 I cannot prove this, but I suspect that 'agency' was a way of 'going beyond' discussions of 'psychology' as formulated by behavioural geographers – unhappily, this move further inhibited the possibility that psychoanalysis would become a resource for (both humanistic and radical) geographers.

6 Compare also Pickles, 1985.

Gold (1980: 14) argued that there was in fact a spectrum between these 'humanisms', and later suggested that 'nothing would contribute more to the future progress of this field of inquiry more than the establishing grounds for a renewal of shared aims and common purposes' (Gold and Goody, 1984: 548). It was already too late: it was clear by the late 1970s that the exponents of various philosophical positions were engaged in very different kinds of projects.

It was common to find behavioural, humanistic and radical geographers stressing the differences between their perspectives and arguing about the relative merits and deficiencies of other approaches. Such arguments tended not only to hide commonalities but also to entrench differences. Thus, some behavioural geographers added the epithet 'analytical' to 'behavioural' apparently in order to strengthen its identification with science. In this vein, Downs and Meyer summed up the differences between analytic behavioural geography and the philosophies of meaning this way: 'for the empiricist, questions of place and space are apprehended from a geometrical, objective stance', while 'the humanistic perspective seeks to reveal those very values, meanings, purposes, and goals, the empiricist eschews' (1978: 61).

While some humanistic geographers recognised the debt to behavioural geography, they were unable to accept its philosophical and methodological presumptions. For Ley (1981a), the most important thing about behavioural geography was its recognition of the importance of the human senses in filtering information. Despite this, he argued that there was a necessary rift between behavioural geography and humanism, because of behavioural geography's unshakeable commitment to an objective and geometrical stance. More particularly, then, Ley could not accept the way in which behavioural geographers became more and more locked into the logics of mathematical modelling, rather than contextual analysis. For him, it was woeful that behavioural geographers should eschew human values, meanings, purposes and goals. While Ley was right to say that behavioural geographers shunned the very things that humanistic geographers thought important, behavioural geography could have been very different.

From the other side of the fence, it is also fitting (and ironic) that behavioural geographers should try to find common ground with humanistic geography by exploring their shared scientific presumptions. Certainly, core humanistic philosophies, such as phenomenology, were themselves versions of science (albeit designed to compete with logical positivism). Given this, versions of humanistic geography were also capable of producing reductionist and deterministic accounts of the relationship between people, environment and behaviour. Nevertheless, humanistic geography more readily stressed that people were creative agents capable of producing fresh meanings and doing startling things.

Meanwhile, radical geographers were arguing that the real determinations of human existence were to be found in asymmetrical power relations. They were much less happy to embed their analyses in the meanings that people gave to their world, since for them the world of appearances never gave a true picture of real social relations: that is, the philosophies of meaning could only ever reinforce the picture of the world which those in power used to disguise their power – and so both logical positivism and humanism became a part of the problem rather than a part of the solution. Certainly, Marxist accounts of social relations relied on the presumption that historical materialism was a superior (if not, true) science. Having implicitly divided the world into science (which uncovered the social structures of power) and ideology (which served to maintain power relations), radical perspectives tended to stress the importance of 'global' socio-spatial relations and the uneven development of social relations between places. However, some Marxists drew on other Marxist traditions which took account of the ways in which classes were 'made' through the relations between classes: such analyses began to take notice of human agency.

In this chapter, I will outline the ways in which humanistic and then Marxist geographers have considered human agency: that is, the capacity of human beings to make their own history (although not necessarily in circumstances of their own choosing). More recently, some humanistic and Marxist geographers have begun a tentative rapprochement, which has involved a reconsideration of the place of people in the world. I have found it intriguing that there is – and maybe has always been – some agreement that there is a dialectic between the individual and the social, or the individual and the environment. The problem that remains is how to conceptualise this 'dialectic':[7] it could be like two sides of a coin, like a two-way street, like a recursive relationship, or – for me – like something much more psychoanalytic. It seems to me that psychoanalysis not only can go between meaning and power, but also locate meaning and power.

HUMANISTIC GEOGRAPHY: MEANING, THE PHENOMENAL ENVIRONMENT AND INTENTIONALITY

According to Lowenthal (1961), the external world did not become a system of facts until it had been organised, first, through the body and, second, through the mind. It was, therefore, inconsistent for analytical geography to presume that everybody responded to an abstract geometrical space which adhered people to a universal logic of spatial behaviour. If it was true that the perception-consciousness system of the mind 'moulded' the phenomenal environment according to belief and value systems ,[8] then 'Man' really ought to be relocated to the centre of the world. While behavioural geography placed a ratomorphic or robomorphic 'Man' into a isotropic, inert spatial

7 Thus, one term (thesis) is defined in opposition to another term (antithesis), where the contradictions within this relationship work themselves out into a third term (synthesis). However, the third term (synthesis) spawns its own internal contradictions. Thus, the third term becomes a first term for another second term: synthesis becomes thesis to its antithesis. And so on – a perpetual recursiveness?

8 As Lynch had argued (see Lynch, 1960: Appendix A).

environment, it was the intention of the humanists to adopt a model of 'Man' in which 'he' was centred, self-reflective and self-conscious, intentional and active.[9] Ley and Samuels put it this way:

> the purpose of the humanist campaign was to put man, in all his reflective capacities, back into the center of things as both a producer and a product of his social world and also to augment the human experience by a more intensive, hence self-conscious reflection, upon the meaning of being human.
>
> (Ley and Samuels, 1978a: 7)

According to Ley and Samuels, spatial science had failed to reflect on the meaning of being human and, therefore, was unable to contribute to the augmentation of human life. At the very outset was a sense that people were both producers and products of their social and physical environment. Thus, the question of meaning and value was placed at the heart of humanistic endeavour. While the 'facts' of behavioural environment were related back either to the hard-wiring of the brain or to overt behaviour in abstract space, Ley (1981a: 214) warned that 'the facts of human geography cannot be viewed independently of a subject whose concerns confer their meaning, a meaning that directs subsequent action. Unlike the natural sciences, then, the social sciences cannot escape the task of interpreting the domain of consciousness and subjectivity'.

Although strikingly similar to the argument made by Kirk (1963), although the humanists (just like the behavioural geographers) focused on overt actions and conscious reflections on meanings, and although humanists (as did the behaviourists) presumed a direct link between meaning and action, the emphasis on meaning was intended to lead to a better, fuller understanding of Man and his place in the world than behavioural geography's.

While most humanistic geographers shared a concern to elaborate human agency and creativity, they were nevertheless divided in their assumptions about the 'centre' of this relationship between people and their lived worlds. Two alternative interpretations can be identified. On the one hand, Anne Buttimer (1976), drawing on phenomenology, saw subjectivity as essentially individual. Although people are individuals, they share their experiences through communication. Thus, the lived world of each individual is limited to a horizon of meaning, communication and understanding. On the other hand, David Ley argued that subjectivity is not essentially or uniquely individual. Instead, the distinctiveness of the individuality derives from the symbolic interactions between individuals: 'life-world is an inter-subjective one of shared meanings, of fellow men with whom he engages in face-to-face relationships' (Ley, 1977: 505). These positions differ on the nature of individuality: the first posits a subject who has an essential character; while

9 I have repeated (here and subsequently) the use of the term 'man' in order to show that this is not a gender neutral way of ordering the world (Lloyd, 1993).

the second presumes that personality develops both through the individual's interactions with others and within the individual's social context. They also diverge on the issue of 'unconsciousness': for the first, meanings are discovered in the 'unconscious' essences of phenomena; while for the second, meanings are found in the conscious interactions between people. Nevertheless, the central issue that both these writers address relates to the origins, nature, circulation and interpretation of intersubjective, shared meanings.

As implied, the meaning of being human was interpreted using two key philosophies:[10] phenomenology (e.g. Buttimer, 1976; Tuan, 1976) and symbolic interactionism (e.g. Ley, 1977, 1978). In the next two sections, I will look at how these geographers built their models of subjectivity from these philosophical resources.

Phenomenology and the lifeworld

For Edmund Husserl, the logical positivism proposed by the Vienna Circle decapitated the world of facts from the world of experiences. In order to disclose the world which lay beyond observable and measurable 'accidents of nature', he developed the philosophy and method of phenomenology. More influential, though, has been the work of Maurice Merleau-Ponty and Alfred Schutz. For geographers drawing on phenomenology, people could be distinguished from other animals by their 'special capacity for thought and reflection' (Tuan, 1976: 267). Nevertheless, people had some kind of primal relationship to nature, place and the world around them. Thus, phenomenology raised special questions about the relationship between people, territory and nature: for example, they 'are able to hold territory as a concept, envisage its shape in the mind's eye, including those parts they cannot currently perceive' and 'the quality of human emotion and thought gives place a range of human meaning inconceivable in the animal world' (Tuan, 1976: 269). The meaning of being human was related not only to people's bodies and minds, but also their experience of place and space (see Tuan, 1977).

Phenomenology provided a people-centred form of knowledge based in human awareness, experience and understanding. More than this, people were not merely passively situated in their environments; they were also active in the creation of meanings, which were bounded in time and space:

> The phenomenological notion of intentionality suggests that each individual is the focus of his own world, yet he may be oblivious of himself as the creative center of that world ... Each knower should recognize himself as an intentional subject, i.e., as a knower who uses words – intended meanings – to render his intentions objective and communicable.
>
> (Buttimer, 1976: 279)

---

10 See Gregory, 1981: 10 and Gregory, 1989a: 360.

While individuals intend to produce certain effects through their words and actions, they are located within circumstances beyond their control. It is argued that a social world, or lifeworld, pre-exists the individual. This lifeworld is characterised by sets of unquestioned meanings and routines which set the frame of reference for individual behaviour; and, because individuals are located within taken-for-granted rules of meaning and behaviour, they can loose sight of themselves as being the creative centre of their world. In this sense, the intentional subject uses words to convey his own meanings and to make his intentions clear to others, but the use of words is governed by 'unconscious' rules and they are shared by an already existing linguistic community. Words seem to traffic overt meanings between people, but they disguise the extent to which they can be personal. From this perspective, the lifeworld may be likened to an ocean where individuals float like icebergs and where the depths and dynamics of the sea can be only vaguely sensed.

An example of phenomenological research would be David Seamon's examination of the 'given-ness' and 'mundaneness' of everyday experience (1979). In a subsequent paper, Seamon discusses the meaning of everyday movements through space, such as collecting mail or taking scissors from a drawer (1980). From the experiences of his students, Seamon illuminates three aspects of everyday movement: first, its habitual nature; second, the importance of the body; and, third, the way in which bodies and places are 'choreographed'. Seamon confirms behavioural geographers' belief that cognitive maps are habitual and repetitive, but argues that they fail to take account of the role of the body. He describes the experience of one student who was driving to his dentist. Instead of turning left as he should have done, the student drove straight on, in a direction which normally took him to his friends' houses. The student explained that this unintentional action was directed '*in the arms*: my arms were turning the wheel ... they were doing it all by themselves, completely in charge of where I was going. The car was halfway through the turn before I came to my senses and realised my mistake' (cited by Seamon, 1980: 154).

Seamon introduces the notion of body-subject to account for this experience of body and space: 'Body-subject is *the inherent capacity of the body to direct behaviors of the person intelligently, and thus function as a special kind of subject which expresses itself in a preconscious way usually described by such words as "automatic", "habitual", "involuntary", and "mechanical"*' (1980: 155).

In analysis, Seamon has moved from a person's experience to the essence of the thing described: from automatic response to 'body-subject'. The idea of 'body-subject' is derived from the intentionality ascribed to the body by Merleau-Ponty (1945). From this perspective, the body has an intelligence and a capacity to act which is outside of conscious control: in this instance, it is in the arms. The body is not a passive mechanism, as assumed by cognitive geographers (citing Tolman, 1948), but capable of its own thinking and acting. Body-subject conveys something of the not-cognitive and not-conscious agency and experience of human beings. While Seamon describes this 'function' as preconscious (and, later, unconscious), the description of the student's experience may well be interpreted psychoanalytically: although the student was consciously intending to go to the dentist, his unconscious wish was to avoid the horror of the dentist's drills and to go somewhere pleasurable, to his friends. His unconscious carried out his (un)conscious wish. Meanwhile Seamon's description of body-subject carries an uncanny resemblance to Freud's body-ego.

There are, nevertheless, significant differences in each approach to the 'psychology' of the body-subject. For Seamon, body-subject is a static essence with its own intelligence, while, for Freud, body-ego is a dynamic concept describing the tension and conflicts within and between the body, the mind and the external world. For Seamon, the notion of a body-subject provides a transparent medium linking cognition and behaviour; whereas, in psychoanalysis, body-ego is in part an effect of multiple internal and external determinations – it is, instead, a concept which explains why direct links between cognition and behaviour constantly fall through the fingers of analysis. Nevertheless, as Seamon intimated,

> consideration of the bodily dimension of environmental behavior indicates that the cognitive perspective is incomplete and needs a thorough rephrasing. Images, subjective impressions, category systems, and cognitive maps may have a partial role in environmental behavior, but we need a better understanding of their relative importance. More than likely, there is some kind of reciprocity between body and mind, habit and wish for change, past and future.
>
> (Seamon, 1980: 157)

Seamon continues his analysis by describing the ways that bodies and places are 'choreographed' through an interweaving of body-ballets, time-space routines and place-ballets. For example, he describes Jane Jacobs' experience of living in Greenwich Village, New York.[11] Jacobs talks about the street as 'an intricate sidewalk ballet' where people move about to relatively strict routines, such as going to school or work (Jacobs, 1961: 52–53). These rituals of everyday life, seen in this way, appear to be choreographed, as stores open and close, as goods are delivered and taken away, as people come and go. This is, for Seamon, the essential experience of the streets of New York: a dynamism of everyday time-space dances, constant movement, a poetics of everyday life – a dynamism which is dependent on the density of people at any one point in time and place (Seamon, 1980: 161). If this is the underlying structure of New York streets,

---

11 It is interesting to observe that David Seamon talks about the experiences of women when he wishes to describe everyday bodily ballets, while Jane Jacobs describes children and men. It is probably too early to presume that the body that comes into vision, that comes to the mind's eye, is already saturated with desire.

then it is one in which no one is mugged, no one is exploited, and nothing is shocking.

As in psychoanalysis, however, the intention of phenomenology is to uncover the hidden meanings of words and actions, their underlying patterns. In this analysis, consideration of the body suggests that there are other dimensions of the mind–body nexus which may determine, or at least condition, spatial understandings, meanings and actions. Moreover, the places through which people move, such as the street, must also be considered if a fuller account of human experience is to be realised. While these are useful observations, the work of other humanistic geographers served to highlight how naïve phenomenological accounts of the social were: for example, Seamon's (1980: 163) description of the essential dynamism of place as 'an environmental synergy in which human and material parts unintentionally foster a larger whole with its own special rhythm and character' could not convey the rhythm and character either of symbolic interaction or of social context .

Symbolic interactionism and social context

Symbolic interactionism is derived mainly from American empirical sociology (that is, the Chicago School) and social psychology (associated with George Herbert Mead). It was based on the idea that human behaviour was founded on shared meanings, meanings that were shared through symbolic exchanges of all kinds, where these exchanges are located in space and time. From this perspective, spatial behaviour and the meaning of things located in space are based on the shared meanings which envelop them. Symbolic interactionism emphasises both the links between symbols of all kinds and the way in which individuals construct, and subsequently maintain, their self-images (see Rock, 1979). These self-images are taken to be symbolic expressions not only of the individual self and the type of shared interaction but also of their place within the boundaries of the social setting. Thus, different senses of self develop through people's selective engagement with the world: that is, through their activities, through their experiences of the world, through their relationships with other people and through location within these interrelationships. There are two aspects of the self (following Mead, 1934): the 'me', who is recognisable to the self as an object; and the 'I', which is the self capable of creative and independent action. This splitting of mental functions into senses of the self is reminiscent of psychoanalytic interpretations of the mind, reflecting Freud's interpretation of the mind through an id, ego and super-ego topology.

Individuals are not just aware of their place in the world, they are also involved in group interactions, all of which are located within social settings; each of these dimensions – subjectivity, intersubjectivity and social context – must be taken into account. Thus, symbolic interactionism suggests that the study of spatial behaviour or spatial patterns must involve the analysis of the interrelationships between the individual, group interactions and the social setting. Moreover, by suggesting that these relationships develop over time, symbolic interactionism also contributes a dynamism to the lifeworld which phenomenology was not able to grasp: 'An individual's behaviour in space and the self which expresses and organizes that behaviour cannot be viewed as being static, but changing via a complex learning process' (Wilson, 1980: 140).

People are constantly presented with new situations, with new people in new places: they desperately try to understand the interactions of others, so that they might themselves behave appropriately. This model not only provided a people-centred alternative for those humanistic geographers who were concerned about the excessive idealism of phenomenology, it also ensured that subjectivity and social context were considered together. The presumption, here, is that human beings organise their emotions, feelings, desires, thoughts and spatial behaviour through selves which, in turn, are defined through their social interactions. Usefully, this breaks the link between 'cognition' and 'behaviour' by inserting multiple, dynamic selves between them. This version of subjectivity is compatible with psychoanalysis, except that psychoanalysis provides a wider and deeper account of these selves and their relationship to one another and the external world. Another important aspect of symbolic interaction is that the self is always located within a situation (see Goffman, 1959). This situation is profoundly geographical: for example, the self who drives to work, or shops, or goes to church.

Wilson (1980) describes the experience of going to church for migrants to inner-city Boston. He argues that the church becomes a focus for the lives of people. Through going to church, the self becomes involved in a highly ritualised set of interactions with others, which is conducted within a particular social and spatial scene. Participation in these symbolic interactions produces strong feelings of intimacy and closeness in people. For Wilson, 'the church becomes a sacred place where order is imposed upon one's pattern of socio-spatial interaction' (1980: 143). These socio-spatial patterns are, it should be noted, defined through symbolic interactions within a particular place. The inner city provides specific opportunities for people to establish interaction within place-based communities, such as church going. That is, the city has no meaning outside of these symbolic interactions. Thus, Ley describes the ways in which the space of inner-city Philadelphia was given meaning and made legible by territorial boundary markers (1974). In particular, he examined the intensity of feelings and attachments which street gangs had for their streets, an intensity which was symbolised through the graffiti which delimited their 'turf'.

The realities of everyday life are negotiated by people in concrete contexts, and so, Ley argues (in contrast to both phenomenology and logical positivism), place is socially constructed. The concept of place is pivotal because it is the contact zone between physical reality, the social context, shared meanings and the self. This argument is based on a particular understanding of the relationship between subjectivity, symbolic meaning and the realities of everyday living. These might be teased out through Ley's discussion of a place: the home.

> The home is the most articulate landscape expression of the self and can reinforce either a positive self-image or, in the case of dreary public housing in an unwanted location, it may sustain an identity of a peripheral and low status member of society with little ability to mould his environment.
>
> There is, then, a reciprocal symbolic interchange between people and places ... Like other commodities, space is engaged not only as a brute fact, but also as a product with symbolic meaning.
>
> (Ley, 1981a: 220)

Ley is arguing that place is a social construction and that there is a fundamental link between place and identity.[12] At the centre of Ley's analysis is not the home, therefore, but the identity of the home dweller. The home is a symbolic landscape for an individual. By using the case of dreary public housing, Ley can show that the home is set within a social context, which sets limits on the ability of the home dweller to change their home. It is through this discussion of social inequality that Ley seeks, first, to avoid a fixation upon consciousness and, second, to deal with context by illuminating the preconditions for, and consequences of, thought and action (see Ley, 1981b: 252). Elsewhere, Ley describes the interrelationship between place and identity as 'reciprocal' or 'dialectical'; arguing that the social world is 'the product of human creativity', but the social world also has 'a certain autonomy', though people's 'autonomy is always contingent' (1978: 52).

Aspects of identity or self develop in relationship to place (people make their homes), but places set a brute limit on what individuals can make of themselves (homes make people). The home is not simply an expression of an individual's identity, it is also constitutive of that identity. Thus, for example, it matters that people in the city are constantly interacting with people they do not know. People produce selves which are capable of acting in a world of abstract, or even alienated, social interactions. It matters that people live in housing estates where they cannot control either the physical or social environment: it produces people who act through particular selves. This account enables four important insights: first, that 'the real' is lost under the veil of images and symbolic meanings; second, that the individual is split into distinct selves or agencies; third, that there is an internal relationship between people and place; and, finally, that people respond in specific ways at different times in different places, partly because they have learnt a 'repertoire' of possible selves.

On the other hand, it is assumed that 'the whole of a person's lived experience involves to a certain extent the attempt to establish some degree

---

12 This relationship between housing, identity and social context is also explored by Duncan and Duncan, 1976, and in Duncan, 1982.

of symmetry between self and external behavior in space' (Wilson, 1980: 145). It is unclear, in this case, what the relationship between the 'me' and the 'I' is: for example, a question arises over 'who' it is that selects 'the self' which is to be presented to others in any given situation.[13] While it is assumed that the mind is split, there is no account of the splitting or of the parts – and therefore there is no account of the diverse relationships that these mental 'functions' might have to the external world. In sum, these ideas cannot distinguish between distinct kinds of internal relationship because the relationship between the 'I' and the 'me', or between selves, is assumed to be unproblematic; nor can they explain asymmetrical relationships between the 'me', the 'I' and the place, because it is assumed that people are constantly trying to harmonise these relationships. It is an understanding of these difficulties which psychoanalysis can contribute to symbolic interactionism.

MARXISM AND POWER TO THE PEOPLE

It is demonstrable that the humanistic project is a sustained inquiry into 'the ways in which human subjects internalise – at a variety of "levels", from the body itself through levels of being, experience and reflection – senses of space, place, environment and landscape that "constitute" these subjects as selves "in the world"' (Chris Philo, personal communication).

Nevertheless, humanism seemed to Marxists to be unable to provide convincing accounts of the power relationships which structure social contexts (see N. Smith, 1979). So, when Gregory's review of Ley and Samuels's progress report for humanistic geography expanded into a full length critique, he was only voicing the frustrations of many radical geographers (Ley and Samuels, 1978b; Gregory, 1981).[14] In this paper, Gregory expressed his irritation with, on the one hand, humanistic geographers' concern with the lived world in its tiniest details; and, on the other, their lack of consideration of the material context within which people live their lives. In humanism, 'the lifeworld' had become both under-bounded (to its tiniest detail) and one-sided (on the side of meaning): paradoxically, covering too much and too little. Gregory suggested that the problems of humanistic geography stemmed from its inability to understand the relationship between structure and agency (1981: 2). Unlike much Marxist geography, Gregory drew on the work of Vidal de la Blache, E. P. Thompson and Anthony Giddens in order to provide a deeper understanding of the duality of structure and agency. He argued that together they offered a way of understanding the 'boundedness' and contingency of practical life (page 5). From these writers, Gregory learns that the real problem is 'to find a model

13 On this problem, see Butler, 1990.

14 For his somewhat subdued response to these criticisms, see Ley, 1982. The irony here is that Ley was, I think, the only humanistic geographer to whom Gregory's criticisms did not really apply.

for the social process which allows an autonomy to social consciousness within a context which, in the final analysis, has always been determined by social being' (Thompson, 1978: 81; cited by Gregory, 1981: 7).

In order to specify the relationship between social being, social consciousness and social context, Gregory draws on Giddens's suggestion that there is an essential 'recursiveness' to social life (1981: 8; following Giddens, 1979). Recursiveness describes the way in which actors reproduce the systems of communication, power and sanction by routinely drawing on existing structures of signification, domination and legitimation. In drawing on these structures, 'the actors involved are displaying some degree of "penetration" of practical life, whether they are able to verbalise their knowledge or not, and that consequently *structure is not a constraint on or a barrier to action but is instead essentially involved in its reproduction*' (Gregory, 1981: 10, emphasis in original).

While people understand their lifeworld to some degree, whether they can articulate their knowledge or not, it is structure that is the medium for the reproduction of practical life. In contrast to much structural Marxism, Gregory set out a position which gave people a kind of contingent autonomy, which was located not just within a diffuse social setting, but within specific social relations, such as 'recursiveness'. In this way, Gregory provided a set of concepts which could be used to understand the relationship between structure and agency, using Giddens's structuration theory (see Giddens, 1984). Meanwhile, other geographers were drawing on similar sources to provide their own accounts of the relationship between subjectivity, practical life and material contexts. It is useful to distinguish between two projects, Thrift's articulation of time geography and Cosgrove and Daniels's radical cultural geography, because each presumes a different relationship between 'structure' and 'agency'.

Determining social action in space and time

Thrift's (1983a) 'On the determination of social action in space and time' remains one of the most comprehensive accounts of personality, socialisation and geography. The paper is set within a time geography perspective (page 28 onwards), though he also draws on writers such as Roy Bhaskar, Pierre Bourdieu, Anthony Giddens and Raymond Williams. Deploying a series of debates in social theory, Thrift highlights a number of substantive issues, most of which are still underdeveloped in geography. In particular, the last section of his paper identifies two major areas for research (page 37 onwards): first, the locale; and second, social action, which is taken to have four interrelated aspects – personality and socialisation, hegemony and knowledge, sociability and community, and conflict and capacity.

Here, I will concentrate on Thrift's conception of personality and socialisation. Basic assumptions about the nature of man are taken from Marx: the human being is an animal which can individuate itself only in the midst of society (page 49); and people make history but not in conditions of their own choosing (page 32). These ideas were cheerfully assumed away by most Marxist geography. Nevertheless, for Thrift, they are taken to present formidable problems (page 37). He identifies two immediate areas of difficulty: first, the process of class formation, class awareness and class consciousness and its relationship to class conflict, ideology and hegemony; and, second, the process of personality formation and the interrelationship of various 'segments' of lived experience, like housing, sexuality, the family, and work. It can be noted that these two difficulties are interrelated: each problem pulls out distinct aspects of what might be called social being, where social being is embedded within differential (social) relationships of power, knowledge and the body.

With respect to space and place, Thrift considers personality and socialisation to be the historical geography of life-path development. For him, the term 'personality' encompasses all aspects of psychology, including subjectivity, identity and individuality (page 43). And, following Giddens (1979), he argues that there are three necessary features of a theory of the person: the unconscious, practical consciousness and discursive consciousness: 'Seen in the contextual sense, personality is a constant process of "internalization" or "interiorization" of social relations along the course of a life path set within civil society' (Thrift, 1983a: 43).

For Thrift, personality is never transhistorical or transgeographical, but 'a continuously negotiated and renegotiated expression of social and economic relations that vary, in other than their most basic form, according to locale and region' and 'a series of sedimented activity-experiences carried forward at the tip of the life path' (page 43). This conception of the personality is easily integrated into structuration theory, where personality is both structured by objective social logic and structuring through resistance to domination. This is a 'dialectical' process, which is termed 'socialisation' and contextualised through the notion of the life-path (page 44). Such an approach is able to recognise the influences of various social forces on any individual; not only race, class and gender, but also time and space. Paradoxically perhaps, for Thrift, the most appropriate level of analysis of these *individual* circumstances is the *population* cohort. This leads him to suggest that the collective experience of the First World War and the Depression, in a regional context, shows markedly different personality formation.[15]

There is much to learn from Thrift's discussion of personality and socialisation. He does, for example, raise the issue of the unconscious and places people in a socially-constructed body (for many, still an innovative idea).[16] However, for Thrift, these are not the most significant issues. He is much more concerned with the dialectical interrelationship between the

15 See also Thrift, 1983b, 1986.

16 The geographical literature on the body is scant. As McDowell and Court (1994), and Longhurst (1995) show, it is confined to articles by feminists, those interested in sexuality and those drawing on Foucault (see, respectively, G. Rose, 1993; Bell and Valentine, 1995; and Driver, 1985).

person and the social. He suggests that, as the person travels through their time-space life path, they 'internalise' and 'interiorise' social relations. This view resonates strongly not only with symbolic interactionism, but also with certain psychoanalytic interpretations. Moreover, like symbolic interactionism and psychoanalysis, Thrift recognises that the person is split: he describes a division into subjectivity, identity and individuality, and another into the unconscious, practical consciousness and discursive consciousness.

The most significant implication of this argument is that questions of human agency need to be framed not only within the determinations (or power relations) of social structure, but also within the material properties of time-space relations, and also within the processes inherent in 'personality'. Whereas symbolic interactionism contributed a dynamism to humanistic accounts of the subject, it could be argued that Thrift's time geography furthers this by emphasising the materiality of social structures and the determinants which surround (but do not extinguish) subjectivity. From my perspective, it can be noted that psychoanalysis aids Thrift's account by suggesting ways in which subjects resist 'socialisation' and by specifying the processes of personality whereby social norms are (or are not) internalised and interiorised. Indeed, psychoanalysis even shows why it is that people feel that they have an interior space.

Marxism, culture and experience

In the same year as Thrift's article, Denis Cosgrove's 'Towards a radical cultural geography' appeared. Although drawing on similar writers, such as Raymond Williams and E. P. Thompson, Cosgrove was attempting to develop a different line of argument from the writings of Gregory and Thrift which were heavily influenced by structuration and time geography. Cosgrove is less concerned with systems, and more with the way that 'human beings experience and transform the natural world as a human world through their direct engagement as reflective beings with its sensuous, material reality' (page 1). This formulation immediately reveals its Marxist origins and its similarities with Gregory's and Thrift's notions of human agency, but it also implies a difference. By introducing the category of experience, Cosgrove is able to suggest that material production and reproduction is 'sustained through codes of communication' (page 1) and that 'if all human production is symbolically constituted we may restate modes of production as modes of symbolic production' (page 8).

The notion of experience performs two invaluable functions in this account: first, it prevents either the person or the social from being collapsed one into the other; and second, it emphasises what might be called the materiality of signs or, alternatively, the political economy of the symbolic. From his perspective, Cosgrove is able to begin to outline the ways in which the subject is located within a dialectic between the material world and the symbolic, or cultural, world. Thus, for Cosgrove, 'the material world is constituted culturally yet remains itself the condition of culture' (page 9).

The implications of this for geography are drawn out in terms of the relationship between landscape, meaning and the social formation:

> Social formations write history in space, and the history of such a formation is the history of the superimposition of forms produced in its landscape through the succession of modes of human production. Since these modes of production are symbolically constituted, place and landscape are immediately endowed with human meaning.
>
> (Cosgrove, 1983: 9)

Cosgrove used an appreciation of the importance of 'the symbolic' to show that material relations are never innocent of human meaning. While Cosgrove was introducing culture to historical materialism, Daniels (1985) was inviting humanists to meet historical materialism. Daniels was aware of the value of humanistic geography, particularly its recovery of the experience of hitherto taken-for-granted notions such as space, place, region and landscape; nevertheless, he argued, the approach would benefit from a historical understanding. But Daniels insisted that these notions were not so much locations, as relationships (pages 145–146). And, drawing on Raymond Williams, this would include material relationships.

Cosgrove's and Daniels's work constitutes a rich vein of ideas. So, their engagement with these ideas enabled them to elaborate the features of an imagined Venice (Cosgrove, 1988, 1989) and Marxism's colonisation of art theory (Daniels, 1989, 1993). Thus, for example, Denis Cosgrove (1982), in his article on John Ruskin's experience of Venice, shows that a significant component of the Venetian myth is the sensuality of that city. Through an interpretation of *The Stones of Venice*, Cosgrove describes Ruskin as

> enjoying the sensuous feast of Venetian architecture ... he wrote of Venice with the passion which betrays more than intellectual interest; his paintings reveal an engagement with the mystery and sensuality of the city, and his fascination with the tides suggest that the organic, pulsing female nature of Venice communicated itself sharply to him.
>
> (Cosgrove, 1982: 164).

The city, the experience of the city and sexuality are intertwined so deeply that, Cosgrove suggests, the city has been feminised in Ruskin's narrative. Thus, Venice 'as a female symbol seems to incorporate many of those ambiguities that psycho-analysis has revealed in the relationship between the child and its mother and between adult male and female as sexual partners' (page 163). For Cosgrove and Daniels, the material world is infused with the pleasures and disgusts of the flesh.

The argument that the material world is constituted culturally is pivotal. For Cosgrove and Daniels, the material world cannot be 'read' except through representations of the world. There is no real world which can be distilled from the cultural, symbolic and personal relations which constitute it. This is a significant point, in part because it suggests that psychoanalytic concepts are always likely to aid the interpretation of the symbolically-constituted world, because the imagination is invested with the ambiguities of desire and disgust, fear and pleasure.

MEANING, POWER AND EXPERIENCE

Despite these promising discussions, the debate over the relationship between structure and agency seemed to dissipate through the 1980s. It is possible that the various factions had decided that they knew what they were talking about and were busy with more pressing problems. It is also possible that the word 'dialectic' was so readily used by everyone that there was actually a broad agreement between the different camps. However it was phrased, there is (or was) a dialectic between structure and agency. After the cross border sniping had died down, the humanists began to stress the experience of material context, while the Marxists began to talk more and more about the experience of modernity. Implicitly, the shared agenda was to locate people's experience within relations of meaning *and* power.[17]

Meaning, experience and context

In 1989, Kobayashi and Mackenzie published a book designed to establish a dialogue between humanism and historical materialism. Although Ley's account of the founding principles of humanistic geography sounded like a recapitulation, certain phrases were introduced to convey the materiality of everyday life. Ley remembers that

> a model of personhood was adopted which upheld the dimension of meaning; human values and experience were integral to a study of people and place. Issues of human agency were central; society consisted of intentional, acting people, and the concepts of subjectivity and intersubjectivity led to the widespread perspective of the social construction of reality ... But action, and particularly the culture-building routines of everyday life, are none the less commonly taken for granted and opaque to actors.
>
> (Ley, 1989: 228–229)

Ley has taken the opportunity to restate the humanist sensibility that meaning and experience are integral to the study of agency and place. The sense that the taken-for-granted world is made up of 'culture-building routines' resonates, interestingly, with arguments being made about social structures by time geographers. Humanistic geography has both emphasised its case and moved on. Accordingly, Ley wants to negate the 'treatment of subjectivity and experience as a virtual fetish, separated from context and material life' (1989: 230). Personhood is situated in the dialectic between the

17 This is evident in much more recent work, see Barnes and Duncan, 1992 and Duncan and Ley, 1993.

individual and their social context, while the lifeworld has an explicit materiality. To this brew, Ley adds a further twist, suggesting that humanism shares a suspicion of 'grand theory' along with postmodernism. Once the poverty of grand theory (by implication Marxist dogma) is recognised, Ley argues, possibilities open up for an integration of humanism and historical materialism. Such an integration, moreover, would restore the relationship between the social milieu and the hermeneutics of everyday communication (page 243). 'An aspiration of humanistic perspectives is to speak the language of human experience, to animate the city and its people, to present popular values as they intersect with the making, remaking and appropriation of place' (Ley, 1989: 227).

While the idea was that humanism would articulate the relationship between meaning, experience and place, the idea that places are made, remade and appropriated introduces a social milieu which is in some senses constituted through power relations. The problem that psychoanalysis finds in this formulation is that 'language' is far from an innocent medium. In clinical experience, the words and actions of patients, their symptoms, have to be interpreted because they simultaneously express and repress experiences. The problem, from another direction, is that language is also a medium of power. Language, moreover, is not innocent of the making, remaking and appropriation of place, as the work of Cosgrove and Daniels shows. Thus, stories about Venice are embedded in, and written through, the desires and fantasies of the narrator. Venice is produced, reproduced and appropriated partly as a myth, partly as a concrete experience, which is never fully conscious, and never completely abstracted from unconscious associations.

Humanistic geography remains committed to the discovery of Man's place in the world, where this world is (more or less) taken for granted. For psychoanalysis, the taken-for-granted is highly problematic: it is not merely a world which happens behind people's backs, nor a world of power relations (as structuralists might have it), but a world in which the internal and external worlds are meshed and separated for the subject. It could be that the world is taken for granted, for example, because it has become mundane, or repressed, or internalised, or externalised. Psychoanalytic understandings of the relationship between the mind, body and external world suggest that meaning is rarely, if ever, transparent and communicable, that the taken-for-granted world might be inaccessible and opaque because it is bound up with the very experiences that language fails to articulate. These ideas suggest that, while shifts in humanistic positions towards the recognition of the materiality of everyday life and of power relations have produced an enhanced understanding of 'the making, remaking and appropriation of place' (page 227), they have yet to come to terms with the depths of subjectivity and intersubjectivity.

Further issues can be elicited from Eyles's suggestion that everyday life is 'the *plausible social context* and *believable personal world* within which we reside. From it, we derive a sense of self, of identity, as living a real and meaningful biography' (Eyles, 1989: 103), and that 'the pattern and meanings of, and reasons for, human actions are structured into and by the societies into which we are born. We both create and are created by society and these processes are played out within the context of everyday life' (Eyles, 1989: 103).

There is an implicit dialectical understanding of the relationship between the individual and the social: thus, a sense of self is derived from a plausible outer world and a believable inner world; and people both create and are created by society. Like Ley, Eyles separates subjectivity and social context and cauterises the flow between them. From a psychoanalytic perspective, the 'dialectic' remains a mystical relationship, which performs in a god-like mysterious way. It is presumed, but never specified. Moreover, it can be quickly observed that not all people live in plausible and believable worlds; indeed, the opposite may be true, that people are constantly trying to make sense of their worlds, trying to make them plausible and believable, and trying to create meaningfulness out of meaninglessness.

Some versions of psychoanalysis see the basic problem of living as dying: in the face of the senselessness of death, we have a real existential crisis which has to be solved somehow – or we die. While this may be pushing the argument too far too fast, psychoanalysis can certainly augment humanistic discussions of experience, first by providing a language which does not presume that subjectivity, intersubjectivity and social context are transparent or accessible, and, second, by suggesting specific ways in which the social is implicated in the constitution of subjectivity and the diverse ways in which people resist their socialisation and set out to redefine their social relations. More than this, psychoanalysis installs a dynamism and ambivalence in the 'psycho-socio-spatial' field which both behavioural geography and humanism want to recover. In short, psychoanalysis can aid humanistic geography's interpretation of 'the social construction of place, landscape, or region, as the interplay between people and context which they both inherit and help redefine' (Ley, 1989: 229).

While humanistic geographers appreciate that society provides people with a sense of identity and a set of culture-building routines, historical materialists have begun to stress the role of agency in the constitution of time-space relationships.

Time, space and social relations

While the humanists have been moving closer to the historical materialists by recognising the presence of the social, the loss of confidence in 'master-narratives' has enabled some historical materialists to move closer to the humanist project. From here, I take a close look at the works of Gregory (1989a, 1989b) because I take these articles to be symptomatic of a broad approach, which includes time geography, structuration and realism.[18] The

18 This proposition can be supported through reference to the work of Thrift (1989a, 1989b) and Sayer (1989).

central concern is with the experience of modernism, but this is a modernism in perpetual flux and forever uncertain: simultaneously exciting and traumatic.

Gregory detects a crisis in modernity (1989a), but sees this juncture both historically and positively. He weaves a new web of social theory and spatial relations in which the postmodern critique of the centred subject is used to articulate a new geography concerned with areal differentiation (1989b). Gregory suggests that there is an inexorable movement towards a 'postmodern' human geography, which is characterised by three key themes: first, the deconstruction of metanarratives; second, the dissolution of the social; and, third, the disorganisation of capitalism. For me, it is important that each of these themes links to what could be called a crisis in western (bourgeois, white, male, modern) subjectivity. From this perspective, the subject can no longer rely on the maintenance of social relations: 'Instead of presupposing that societies are totalities with clear-cut boundaries, it becomes necessary to show how social relations are stretched across varying spans of time and space' (1989a: 354).

The subjects may derive their identities from the world around, but that world is no longer stable or fixed: social positions are mobile, up and down. Giddens's work is worthwhile because for him the theory of structuration is a sustained attempt 'to dismantle the dualism between agency and structure and replace it with a duality' (Gregory, 1989a: 354). For Gregory, the notion of duality is an important advance not only because it presents the structure–agency relationship, but also because it connects this relationship to time and space. Thus, the notion of duality suggests an interdependence between practical life and time-space relationships, where

> neither society, nor individuals are assumed to exert a greater influence on events than the other. The relationship between agency and structure in time and space is treated similarly; whilst temporal and spatial organisation limit individual action, they are, at the same time, the creations of history, society and individual action. Again, each exerts a determining influence on the other but this is again of equal weight.
>
> (Gregson, 1986: 185)

Linking space and time to the structure–agency relationship is a step forward, as is noticing that time and space are interdependent. However, while it may seem like an advance to presume equality and equivalence between mutually-determining opposite terms, it is also possible that things do not happen this way. There is something familiar about the presumption of a 'dialectic', which I think cannot be presumed (see above). Nevertheless, what is important about Gregory's analysis, for me, is that he is prepared to assert certain basic propositions about the experience of modernity.

These propositions are deceptively simple: first, that time-space relations are constitutive of society; second, that time-space relations are constitutive of knowledge; and finally, therefore, that 'time-space distanciation is structurally implicated in the time-space constitution of knowledgeable and capable human subjects', where time-space distanciation is the stretching of social relations across time and space (Gregory, 1989a: 376). The problem, as Gregory sees it, is how to understand 'the modalities through which power enters into the constitution of human subjects' (1989a: 377). Gregory now turns to the constitution of subjectivity. The treatment is revealing. For Gregory, Giddens's understanding of subjectivity is flawed.

> While [Giddens] evidently does not conceive of the human subject as somehow 'preformed', he undoubtedly offers an account of the constitution of human subjects which is, at bottom, ahistorical. He treats subjectivity in strictly developmental terms, drawing on the ideas of Erikson, Freud, Lacan and others to establish the transformation of the body into an instrument of acting-in-the-world, and subsequently using Goffman, Hägerstrand, and others to emphasize the significance of time–space routinization for sustaining the stratification of personality.
>
> (1989a: 377)

Gregory is suggesting that Giddens's reliance on psychoanalytic accounts of psychological development produces a universal and timeless (and therefore spaceless) account of the subject. He is, rightly I believe, objecting to any notions of subjectivity which imply that people have the same psychological make-up no matter what the (historical) circumstances, no matter where they are. Similarly, Gregory criticises Giddens's acceptance of, and emphasis on, time-space routinisation as a universal basis for the 'stratification' of personality. Gregory's knife cuts several ways: into structuration, into psychoanalysis, into symbolic interactionism and into time geography. Nevertheless, his target is the same: universal presumptions about the constitution of the subject. Habermas's account is preferred because 'his theory of social evolution, whatever its demerits, is at least designed to draw attention to the different ways in which human subjects are constituted in different types of society. It is this insight which is missing from structuration theory' (1989a: 377).

Gregory argues that Habermas's conception of the human subject is nevertheless insufficient because of his denial of the spatiality of social life. Thus, neither Habermas nor Giddens offer a complete solution to the problem of the constitution of the human subject, but together they might. Gregory proffers a historical geography of the person which is strategically complete because it would take account of both time-space routinisation and time-space distanciation in the time-space constitution of the human subject.

For me, this is an important argument. However, I would add that psychoanalysis can also contribute to Gregory's examination of the 'time-space constitution of the subject'. It would be noted, for example, that routinisation and distanciation are part and parcel of the constitution of subjects, but that these are also constituted by subjects, and are not the only 'time-space' relations which are implicated in subject formation. Indeed, they might not even be the most fundamental ones. Instead, it might be possible to develop arguments around aspects of repetition, repression, object relations and simultaneously real, imaginary and symbolic spatialities – none of which presume a universal subject or an inevitable trajectory for psychological development. I might also comment that psychoanalytic discourses are constructed out of the pain and torment of individuals and, while this suggests certain general circumstances of pleasure and pain, one could hardly presuppose that psychological development is universal (see also Chapter 4 below). Unlike Gregory, I would have abandoned both Giddens's and Habermas's account of social and individual development because neither can specify the 'dialectic' or 'duality' of structure and agency, partly because they operate with static notions both of subjectivity and of language use.

Gregory's work displays a commitment to rethinking the relationship between the individual and the social, between meaning and power, and between time and space. It is my contention that psychoanalysis can help this project, but there seems to be a problem in the way in which structure and agency are conceived. The problem seems to lie in the dialectical mode of understanding: not necessarily in the notion of a dialectic, but in its hidden effects on the 'things' it conjoins, contrasts and counter-poses.

The dialectics of human agency

Andrew Sayer has summed up the relationship between the subject, space and society this way: 'what you are depends not just on what you have, together with how you conceive yourself, but on how others relate to you, on what they understand you to be and themselves to be' and 'to a considerable extent people have to adopt meanings, roles and identities which pre-exist them' (1989: 211 and 213 respectively). A variety of terms have been used to describe this relationship between, on the one hand, intentionality, subjectivity or agency and, on the other, the taken-for-granted, social context or structure: for example, reciprocal, recursive or mutually determining. These concepts are meant to convey, I think, three things: first, that the relationship between the individual and the social is equal; second, that it is dynamic: and, finally, that it is two-way. This relationship is, by any other name, a dialectic.

Whatever the phrase, a pair of terms face one another, as opposites, in an intricate dance, which may have an end or not, where the partners may tread on each other's toes or not, they may even love each other, depending on how the dance is interpreted. The pairs are, however, inseparable, constantly moving and always engaged in what the other is doing. Dancing may not be the most suitable metaphor for a dialectic,[19] but it does highlight a growing sense that you cannot understand 'the dance of life and death' by just looking

19 There is the world of difference between the Gay Gordons, the last tango, ballet and disco – for example – thinking about their social *and* spatial settings, their historical development, their performativity, and who is in control.

at one partner and ignoring what the other one is doing. It also shows how the dialectic joins two opposing terms together in a determined choreography: creativity is locked in a death dance with the taken-for-granted, where agency is the mirror image of structure – subjectivity is what social context is not, and vice versa.

To follow the dance metaphor for a moment: it could be that you would want to look at the rules of the dance or what the dance means, but it may be that the partners are in it for different things, that they are not always in step and that their minds may be on other things. There could be many ways in which the pairs are – and are not – engaged in the same shuffle. It is this that is worth exploring. For me, it requires more than both humanism and historical materialism to show how the human subject is (re)created, with(in) forms of power, even while it is necessary to face the dialectic of structure and human agency (see Kobayashi and Mackenzie, 1989: 10).

The problem is that – once social life has been partitioned into structure and agency, structure detached from agency and structure insulated from agency – structure is seen as taken-for-granted, common and external, while agency is seen as self-conscious, individual and internal. In using the idea of the dialectic, this problem is at once invoked and swept under the carpet, a magic carpet which keeps the dancers on the (same) dance floor. From this perspective, the movement and noise within, between and beyond the dance is stilled and silenced. For me, it is this (stilled) movement and (silenced) noise which psychoanalysis watches and listens to – not as a voyeur, or eavesdropper, but as a participant. The question, here, is how to specify the dialectic, not only in terms of the contents of its binary opposites, but also in terms of a set of differential relationships, including the points where they break down.

CONCLUSION: SUBJECT FORMATION, CIRCUMSTANCES AND SPACE

Pressing problems begin to arise once the processes of subject formation, of people's relationship to others and of becoming self aware, are called into question. These issues cannot be localised, they bleed from every pore of the body of geography. From a humanistic perspective, Eyles has suggested that

> the study of everyday life must also recognise self as the being which conceptualises and acts in everyday life. Self is reflexively aware, but a significant part of identity is based on the presence of others. We experience ourselves simultaneously as subjective sources of projects and as objective reflections and reactions of others. We must mutually construct each others' lives and this unquestioned construction is the basis of sociality, which itself occurs as social interaction. Indeed we define ourselves in terms of the accumulated history and anticipated future of interactions that make up our lives.
>
> (Eyles, 1989: 115–116)

Every sentence is an open wound: what precisely are the relationships between 'the self' and 'being' (for example, are they co-extensive as Eyles suggests, or are there tensions between the self and other aspects of being?), between the self and self-awareness (is the self, for example, only related to that part of the self which can treat the self as an object of reflection and what about the parts of ourselves that we are unaware of?), between identity and the presence of others (what happens if we, if only in part, misrecognise others?), between the subject's actions and the actions of others (what if actions have unintended consequences and meanings?), between actions and social norms (are people following rules even when they think they're not?) and so on? These questions are, I think, related. They also intrigue me. As Thrift argues,

> all manner of problems present themselves – the explanation of practical reason, how people fashion accounts, the nature of the self. These problems are intensely geographical. People are socialized in localized contexts (although the institutions of socialization are now rarely local) and the exigencies of these contexts produce different people with different capacities to think, to co-operate, to dominate, and to resist.
>
> (Thrift, 1989a: 152–153)

The heart of what has been at stake in this chapter is found in Marx's oft-repeated aphorism: 'people make history, but not in circumstances of their own choosing' (Marx, 1852: 146). The problem lies in the precise relationship between 'people', 'history', 'circumstances' and 'choosing'. Thrift shows that, in trying to understand the position of 'people' within 'history', social theory has usually decided to resolve the issue either on the side of 'history' or on the side of 'people'. To simplify greatly: on the side of history, it is argued that circumstances by and large determine what people choose to do – from this position, it is a short step to believe that circumstances determine what people do and that people are trapped within the logic of the social structure (whether this is determined by capitalism or patriarchy or some other relation of power); on the side of 'people', it is argued that individuals make their own choices, though limited by certain circumstances – from here, it is a short step to believe that people are completely free to choose what to do, without constraint on their actions.

These positions carry implications for understanding the relationship between the subject, space and the social. *From the perspective of structural determinations*, the subject has no meaning outside a system of social relations: thus, the individual only has meaning as, for example, 'labour-power' or 'male', or subjectivity only has meaning in relationship to 'class consciousness' or 'masculine', for example. Whatever the theory of the dominant relation of power, it is this that fills the empty vessel of the subject and gives it meaning and identity. Outside the dominant system – whether it be capitalism, patriarchy, or something else – the subject is assumed to be nothing. The challenge, then, is to change the system. *From the perspective of the intentional self*, the subject has their own internal meaning and identity, though this is commonly taken to be hidden under a great depth of shared ideas. The individual's experiences are seen as central to the disclosure of their (true) meaning. The subject can mean anything that an individual takes it to mean, taking on different qualities at different times: thus, it could be experienced as masculine, skinny, white, young and so on, depending on the way the subject is coded in a social setting and the way that setting is decoded and recoded. Because the essence of the subject is found either in their own inherent qualities or in their intersubjective experiences, received meanings are open to challenge.

Conceptualising the relationship between 'history' and 'people' as a dialectic was intended to enable a cartography of the subject in which, on the one hand, social structures at least set the parameters within which people behave, but, on the other hand, at most set the rules for 'allowed', 'prohibited' and 'enabled' thoughts and actions. Either way, the problem which Thrift raises has something to do with the relationship between socialisation and space. So, for Thrift, 'human agency must be seen for what it is, a continuous flow of conduct through time and space constantly interpellating social structure' (1983a: 31). People are socialised in place, in specific circumstances:

> Through the processes of socialization, the extant physical environment, and so on, individuals draw upon social structure. But at each moment they do this they must also reconstitute that structure through the production or the reproduction of the conditions of production and reproduction. They therefore have the possibility, as, in some sense, capable and knowing agents, of reconstituting or even transforming that structure.
>
> (Thrift, 1983a: 29)

If I have understood the project, then, the problem is to understand not only the dialectic between the subject and the social, but also the various ways in which space is implicated in this dialectic. The difficulty I have identified is that the dialectic is itself part of the problem of understanding the relationship between the subject, the social and space. I have hinted that the subject is not one thing, neither is the social and nor is space, that if there are dialectics then they may be of different kinds and orders, and that these relationships may be multiple, interrelated and dynamic. I suggest that a psychoanalytic interpretation of these relationships would aid an alternative conception of environment, mind and behaviour, of agency and of structure. I can point to the production within psychoanalysis of two distinct, but co-existing and related, topographies of the mind (neither of which can be understood, except spatially), to the various ways in which subjectivity is installed spatially in psychoanalytic narratives of childhood development, to the ways in which space is understood and manipulated by the subject and, finally, to psychoanalysis's preoccupation with the break points of socialisation, with sites of resistance, and its continual encounter with places beyond understanding. With psychoanalytic maps, it might be possible – warily – to chart the *terrae incognitae* of hearts and minds, of meaning, identity and power.

CONCLUSION TO PART I

This part of the book has travelled a long way. As with any journey, it has covered only a small part of the countryside and met only a few people. Nevertheless, these have been significant places and people, those involved in the production of knowledge about the relationship between subjectivity, space and the social. In this conclusion, I intend, very briefly, to review 'the problem' and to suggest how psychoanalysis might contribute. I will begin with the way the problem was formulated by behavioural geographers, then move on to a discussion of meaning, power and identity. I end this conclusion by outlining the kinds of psychoanalytic argument I am not interested in engaging with – and there are hints of the kinds of themes I will be developing, but these are properly outlined in the Introduction to Part II.

Let me begin with behavioural geography. Of behavioural geography, Walmsley and Lewis argued that: 'What seems to be needed is an approach that is individualistic, action-oriented, and which allows man a modicum of free will and a degree of latitude in interpreting, and ascribing meaning to, the environment' (1984: 39).

Behavioural geography, like humanistic geography and variations of western Marxism, wanted to see things from the point of view of people. The constitution of the individual now becomes the central most important question. Walmsley and Lewis's use of the term man should set alarm bells ringing over the notion of the individual being assumed in their statement: their man is mapped through the co-ordinates of 'free will', 'interpretation' and 'creativity'. Such ideas can be linked to Enlightenment thinking, which has for some time been known not to guarantee enlightenment.[1] Leaving this to one side, Walmsley and Lewis's approach may map the (western male) individual into the familiar territory of free will and creativity, but they are unsure of where the individual is: their tentative qualifiers – 'a modicum of' and 'a degree of' – speak of the difficulty of understanding the relationship between the individual, overt spatial behaviour and the external world.

The question remains this: how are the relationships between the internal world and the external world to be conceived? An early approach to this

1 For examples of this critique, see Horkheimer and Adorno, 1944 and Lloyd, 1993.

question started by positing the simplest case – the individual at rest – and thinking about what happens next. Golledge, Brown and Williamson pictured the situation this way:

> Imagine an individual in an unmotivated state (i.e. before he is stimulated to make a decision). He can be described by a set of personal functional variables (such as his innate mental and physical abilities, and his value system); a set of personal structural variables (such as his age, education, income, occupation), and a set of existence variables (such as location and orientation with respect to elements of the physical world) . . . Next, the individual is stimulated by being exposed to a conscious or unconscious drive or cue, and must make some decision.
>
> (Golledge, Brown and Williamson, 1972: 63)

There are some serious issues here; each sentence carries significant assumptions. Let me tell a short story. I have, on several occasions, been around a special care baby unit in a Bristol hospital. These babies ought still to be in their mothers' wombs:[2] instead, the womb has been replaced by medical staff and machinery. It is clear from their movements that these tiny sleeping cyborgs dream – what they dream is anybody's guess, but one thing is for sure, even at this you-should-be-in-the-womb age, they are never in an unmotivated state. What this tale highlights is that Golledge, Brown and Williamson are working with a very specific notion of the 'decision'. Decisions are conscious and communicable (they can be described) – they are, by extension, adult and nothing to do with (cyborg) babies. Psychoanalysis, it can be noted, works with linguistic, non-linguistic and prelinguistic material and presumptions. It also shows that tiny babies are already (and always?) living an unconscious life: the physical world does not happen to them as if from the outside, as Golledge, Brown and Williamson imply (I pursue this point about the unconscious below).

Golledge, Brown and Williamson suggest that this unmotivated (dead?) man is the sum of his situated variables – of which there are three classes: personal functional, personal structural and existence. In their classification of variables a number of disparate elements are conflated and segregated. It is possible to argue about these, but the point here is that though these variables may be components of a description, they do not tell you either what kind of person the individual is or how they will react in any given circumstance.

The lifeless state is only a starting condition, however, so next this inanimate Frankenstein's monster is sparked into life by 'a conscious or unconscious drive or cue'. Then, the monster has to make a decision: at risk of stating the obvious, what has not been theorised in this account is the

---

2 This is not a moral 'ought', but a nine month 'ought'.

'unconscious drive'.[3] It is possible that Golledge, Brown and Williamson are referring to the kind of instincts that led to the original monster's (and Frankenstein's) undoing: innate impulses which determine behaviour, whatever the circumstances. On the other hand, these could be drives in the Freudian sense of a general energy which is given form within the child, within its encounters with others and within the sanctions of culture in specific (rarely the same) directions. From a psychoanalytic perspective, drives are a dynamic process involving a source (an internal bodily stimulus), an aim (to deal with the tension derived from the stimulus) and an object (thanks to which the aim might be achieved). How the drives are dealt with, and the kinds of objects to which they become directed, relate to the interaction between the psyche, the body and the external world (in all its forms). According to Freud, there are two basic sets of drives, Life and Death, which are in-conflict, indefinite, and even mythical, energies, rather than being an either/or grid of behavioural determinations. I would guess though that Golledge, Brown and Williamson are not drawing on Freud to support their arguments.

The unconscious has always been a stumbling block for behavioural geography, a continual presence which is pushed to one side, out of the way, under the carpet. Behavioural geography has shared this attitude with humanistic and radical geography. Nevertheless, they do not share a theory of consciousness. Differences between analytic behavioural geography and humanism can best be seen in their attitudes toward the mind: 'in the empiricist conception, mind means brain in the psychological sense. For the humanist, mind is much closer to the notion of spirit (Geist), of mind-in-the-world' (Downs and Meyer, 1978: 61). For Downs and Meyer, this disagreement is grounded in the researcher's assumptions about 'what can be known of the world, including physical reality, consciousness of that reality, and also awareness of that consciousness' (1978: 67). The question of 'awareness' is at the heart of this problem.

Humanistic and radical geography attempted to get beyond the conscious, observable, known world by drawing on analysis of meaning and power. While both sides agreed that there was a dialectic between 'the creation of meaning' and 'structures of power', they mapped overt spatial behaviour according to the circulation of meaning or the determinants of structure. Despite this, the agency/structure dialectic has not completed the map of the subject,[4] and the debate has been recast onto the terrain of language or more properly of discourse. One outstanding problem with the way the agency–structure dichotomy operated was that it still seemed unable to interrogate 'everyday life' as simultaneously real, imaginary and symbolic, as

3 I am not sure whether this sentence reads unconscious drive or unconscious cue, or unconscious drive or some other kind of cue, so I will just respond to the notion of a drive.

4 A wide range of alternative models of the subject have been introduced to geography recently, see Bell and Valentine, 1995 and Pile and Thrift, 1995a.

simultaneously about meaning, power and subjectivity. The assumption was that terms such as the subject, the body or the city had identifiable meanings informed by their location within the dialectic between agency and structure. This assumption was challenged in post-structuralist thought.

By concentrating on discourse simultaneously as an identifiable situated social practice and as a relationship involving power and knowledge, it was possible to argue that maps of subjectivity and space were constituted by the practices of everyday life that they seemingly described. Thus, institutions such as the madhouse, prisons, hospitals and universities, rather than containing particular subjects, actually and actively create them: thus, asylums create the insane, universities create students.[5] Insanity and academia are inconceivable outside of the institutions that give them meaning. The agency–structure dialectic had been circumscribed: discourse was neither structure nor agency and both structure and agency. From this perspective, the individual becomes a longitude and latitude of various power-infused discursive positions, but where they are not a passive medium on which cultural meanings are merely inscribed; they are neither an essence nor a free-floating set of attributes.

Aware of the discursive production of subjectivity and the facts of life, Elspeth Probyn proposes that the self

> is a doubled entity: it is involved in the ways in which we go about our everyday lives, and it puts into motion a mode of theory that problematizes the material conditions of those practices. Unlike the chickens which are presumably sexed one way or the other, once and for all, a gendered self is constantly reproduced within the changing mutations of difference. While its sex is known, the ways in which it is constantly re-gendered are never fixed or stable.[6]
>
> (Probyn, 1993: 1)

The individual is not a double entity – located within the agency–structure dialectic, but a doubled entity: multiplied, dynamic, participating and determined – and enduring doubling which constantly produces another place, the site of constantly changing mutations of difference, at once stable and dislocated, at once fixed and changing. The subject is never in one place. Thus, the focus of an analysis of the subject has changed from identifying their location on the continuum between structural and personal determination to looking at the ways in which subjectivity is reproduced in time and space, within the interrelating relations of meaning, power and identity. From this perspective, for example, unlike the chickens, the truth of sex is 'performatively produced and compelled by the regulatory practices of

5 It should be clear that the debt here is to Foucault (for example, 1961, 1966).

6 To explain 'unlike the chickens', Probyn refers to the practice of sexing chickens on a Welsh hill farm as either cocks or hens. There are no mutations of difference, no 'third' terms for the bodies and sexualities of chickens.

gender coherence' (Butler, 1990: 24). The cartography of subjectivity is, in this account, reproduced both through situated practices which make gender legible and through power-laden regulations which legislate gender norms: 'Gender is the repeated stylization of the body, a set of repeated acts within a highly rigid frame that congeal over time to produce the appearance of substance, of a natural sort of being' (Butler, 1990: 33).

Men might not like to think that wearing Marks and Spencer's shirts and Levi jeans is a stylisation of the body which makes their gender identity coherent and legible, but it nevertheless is. Subjectivity becomes a performance – a masquerade – which does not have an essence, but seems to, although not thinking about it. People are embedded in constellations of relationships, which involve feelings and responses derived from many interconnecting and disconnecting sources. In everyday lives, people assume themselves to be the kind of people they are. They react to people on the basis of the kinds of people they seem to be. Intricate, unfolding interactions take place on a daily basis: internally, in relation to others, within particular circumstances, as they seem to be. The body language of others is read, their words and expressions interpreted, as are the feelings that are provoked by these circumstances and the moods that people are in. In all this, it is commonly presumed that people know their own minds, that they can express their thoughts and feelings, and that they act according to their rationalisations.

> You feel sure that you are informed of all that goes on in your mind if it is of any importance at all, because in that case, you believe, your consciousness gives you news of it. And if you have had no information of something in your mind you confidently assume that it does not exist there. Indeed, you go so far as to regard what is 'mental' as identical with what is 'conscious' – that is, with what is known to you – in spite of the most obvious experience that a great deal more must constantly be going on in your mind than can be known to your consciousness. Come, let yourself be taught something on this point! ... You behave like an absolute ruler who is content with the information supplied him by his highest officials and never goes among the people to hear their voice. Turn your eyes inward, look into your own depths, learn first to know yourself! Then you will understand why you were bound to fall ill; and perhaps you will avoid falling ill in the future.
>
> (Freud, 1917a: 142–143; cited by Ricoeur, 1974: 152)

You may feel sure that behavioural, humanistic and radical geography can inform you of the cognitive maps that people have of the city, of the time-space ballets of the choreographed body, of the political economy of the uneven development of urban space, but their emphasis on visibility, legibility and consciousness might not necessarily inspire confidence that all the bases have been covered. Indeed, it may be that the 'real' determinants of overt spatial behaviour are not only covert but also spatial in other ways. What is of importance here is that the mental and consciousness are not synonymous. There is an unconscious, which reveals itself in some obvious experiences, famously such as dreams and slips of the tongue (see Freud, 1900, 1901). Have you ever dreamt; has anyone ever told you that you just mispronounced a word?

The word unconscious is not Freud's and the concept can be used to convey meanings Freud did not intend. It does not, for example, mean not-conscious as in yet-to-be-known or taken-for-granted; nor does it imply short circuits in the brain or well-worn neural pathways. In Freud's work, the unconscious is another scene, a parallel process which works by its own logic; it uses its own language, signs and symbols, makes its own connections; it is born out of prohibitions, repressions and taboos – all of which are nested in the psycho-social-spatial field of everyday life. It is a zone of primal exclusion, it is a wound, but it knows no Law, no negation, everything is permissible – it is mythic and magical – yet it is as real as anything else.

Freud warns: you may act like an absolute ruler but, never mind the intermediaries, you must listen to the people – what tongues are they talking in, where are they talking from – otherwise how are you to understand your place in the world? And psychoanalysis calls into question the neat homology between identity, self-awareness and consciousness. Is it really possible to 'know yourself'? If the injunction is to know yourself, then it is only a first, fumbling step on an impossible journey. A journey into a mythic field – into stories about legends: Narcissus and Oedipus. Before developing my argument through telling Freudian tales (in Chapter 4), I should first put down some caveats (I have already dropped some hints) about what is to follow.

Psychoanalysis is in itself an enormous literature and it has had a long-standing and widespread influence on the human and social sciences – even a list of representative works would run to several pages. So, it is useful to say what I am not going to be looking for in psychoanalysis before moving onto the next part of this book. The path I follow through psychoanalysis is (to use Butler's expression) 'rigorously unorthodox' and designed to avoid one of its well-known pitfalls: that is, its complicity in the very power relations it uncovers and names.

I think it would be possible to take certain Freudian mythologies and apply them to bodies and cities. For example, I might say that the Oedipus complex is universal – it tells of castration anxiety and repressed love. I could add, in all confidence, that tall erections (ho, ho) in the city – such as towers or monuments – are men's response to, and expression of, this formative childhood anxiety. I might say that women never fully abandon their narcissism and that they derive their sense of worth from their ability to reflect the desire of men. I could add that it is women who love to shop 'till they drop, that their sense of self is embedded in the experience of shopping, which covers their fear that they are not desirable in a cloak of desirable goods.[7] I may continue that shop windows act simultaneously as a peephole

---

7 Not such a ludicrous idea as it might at first sight seem, see Bowlby, 1993.

and mirror on desire. In both cases, the body and the city are reflected one in the other and this is an inherently gendered story. And I've psychoanalysed these situations. But I have failed to provide either an adequate or a convincing account of the real, imaginary and symbolic spatialities in these relationships. So, I am not looking to psychoanalysis to provide universal, fixed or static 'truths' about people and the world. Indeed, I think that such a project would run against the grain of psychoanalytic thinking.

Psychoanalysis is, after all, a spatial discipline.

Part II

SPACES OF THE SUBJECT

INTRODUCTION TO PART II
geographers and psychoanalysis

Psychoanalysis is a controversial account of mental life and a troublesome form of knowledge. Unsurprisingly, therefore, there are no accepted psychoanalytic concepts which can be easily transposed into, superimposed onto, or mapped alongside, geography – regardless of the kind of geography. In Part I, I demonstrated that various forms of geography have avoided a layer of explanation and understanding, the unconscious, that psychoanalysis is particularly well informed about. Nevertheless, psychoanalysis has been left aside by geographers. It is easy to claim that psychoanalysis has been systematically misrepresented, but I would prefer to suggest that particular aspects of psychoanalysis have been selected and presented as if they were symptomatic of the whole approach. Until very recently, then, only specific characters from psychoanalysis, wearing grotesque masks, have been allowed to take the stage and they have been booed off. From stage left, however, psychoanalysis has appeared in other, different guises. The audience waits – tentatively and suspiciously.

Up to the late 1980s, the most fulsome descriptions of psychoanalysis were to be found in Roger Downs and David Stea's introduction to *Image and Environment* and John Gold's *Introduction to Behavioural Geography*; though each of these treatments amount to no more than one page each. The most thorough review is found in Michael Smith's *The City and Social Theory*, in his assessment of the (ir)relevance of Freud's ideas to theories of the city. I will consider these responses in the first part of this introduction.

Lately, in the last five years or so, references to psychoanalysis (in particular, Lacan) have spread like measles over the body of human geography. Spots can be found in studies of landscape, nature and environmental history, in arguments over the bones of modernism and postmodernism, in cultural geography and in geopolitical geography. Like measles, this disease probably needs two weeks of isolation, a good rest and a bath of cold porridge. Joking aside, psychoanalysis remains a marginal resource for geographers – it seems to be an unnecessary, unpleasant experience, just like measles. Sustained engagements with psychoanalysis can be found, however, in the work of five geographers: in more or less chronological order, Gunnar Olsson, Jacquie Burgess and David Sibley, Gillian Rose and Heidi Nast.

In the second section of this introduction, I will review the work of David Sibley and Gillian Rose. This is not to dismiss the work of Olsson, Burgess and Nast – far from it – but there are local reasons for putting their work to one side. Gunnar Olsson's writings have been an attempt to uncover the rules and practices of the language game and the ways in which power inveigles itself within language; throughout this project he has used the work of Lacan to describe the phallocentrism of signification (see Olsson, 1980, 1991). While Olsson draws on a psychoanalyst, Lacan, his work is not about psychodynamics and so does not fit into the line of argument of this book. Jacquie Burgess's research has consistently used principles derived from the practice of group therapy outlined by Foulkes (see Burgess *et al.*, 1988a, 1988b; Foulkes and Anthony, 1957), but I have discussed this work elsewhere (Pile, 1991). Heidi Nast is producing an extraordinarily creative stream of ideas; however, this work is mostly unpublished, though I will draw on Blum and Nast's critique of Lefebvre in the second half of this book (Blum and Nast, forthcoming).

The deepest and most thorough elaborations of the relationship between psychoanalysis and geography are found in the writings of David Sibley and Gillian Rose. David Sibley has interwoven Julia Kristeva's notion of abjection and Melanie Klein's work on children to talk about the segregation of space and the spaces of childhood (1988, 1995a, 1995b) and the post-Lacanian psychoanalytic feminisms of Luce Irigaray and Teresa de Lauretis have been deployed by Gillian Rose (1993, 1995). I will deal with their work in turn below, but first let's see how geography was introduced to psychoanalysis.

PSYCHOANALYSIS FOR GEOGRAPHERS?

For Downs and Stea, psychoanalysis was the exact reverse of rational economic man:

> Psychoanalytic man, as delineated by Freud and Jung, was totally non-rational. His adult behaviour was determined in large part by the (probably unconscious) resolution of psychological conflicts experienced earlier in life, and was influenced by biologically transmitted traces of earlier experiences in human evolution ('collective unconscious' or 'racial memory'). External factors were assumed to play a small role in adult patterns of decision-making: social influence was secondary, and environmental influence negligible.
>
> (Downs and Stea, 1973b: 3)

In psychoanalysis, adult behaviour cannot be understood outside of the ways in which people (in part unconsciously) resolve 'psychological conflicts experienced earlier in life', as Downs and Stea note. However, their description of psychoanalytic man (sic) is wide of every other mark: for Freud, human psychology is not 'totally non-rational'; for Freud, there is no biologically transmitted 'collective unconscious' or 'racial memory' which bears the traces of experiences throughout human evolution (in any case, these are more in the style of a Jungian analysis);[1] and external factors are not presumed to play a small role in adult patterns of behaviour. Indeed, this quote may say more about the way that Downs and Stea wish to conceptualise 'decision-making' as rational and largely determined by external factors. Their caricature of psychoanalysis produces a mirror image of rational economic man: namely, non-rational psychoanalytic man. Injury is added to insult: both rational and non-rational man are rejected. By doing this, Downs and Stea not only marginalise the non-rational, but also childhood feelings and mental processes.

In a more sophisticated appraisal, Gold suggested that Freud showed that 'behaviour was largely decided by drives acquired in early childhood, and thereby stressed the previously neglected role of subconscious motivation' (Gold, 1980: 12). Although Gold goes along with the idea that behaviour is not largely determined by conscious rational behaviour, he does not suggest that drives are 'non-rational' or that this model is 'totally non-rational'. Moreover, Gold believed, Freud created a dynamic model of the mind (unlike a static 'non-rational' mind): 'According to Freud, personality rested on the dynamic interaction of three mental elements: the id – the storehouse of subconscious sexual or aggressive impulses; the superego – the conscience; and the ego – the rational self which tries to satisfy the demands of both id and superego' (1980: 12).

Although Gold values the id/ego/superego distinction and recognises the dynamic relationship between them, he sees regressive tendencies in psychoanalysis which 'portrays a non-rational decision-maker, whose behaviour reflects the resolution of inner conflicts [which] count for more than external environmental conditions' (1980: 12).

Somehow, in a quick act of forgetting, the ego – which Gold calls the rational self – has been replaced by 'a non-rational decision-maker'. By ignoring the id and the super-ego, Gold has somehow fallen back into Downs and Stea's parody of Freud's beliefs. A trap has been sprung and Gold's argument is caught by it. He now outlines three problems with psychoanalysis: 'There is little doubt that it oversimplified personality structure, undervalued the importance of intelligence and learning, and relied upon empirical evidence derived from studies of mentally sick or distorted subjects' (Gold, 1980: 13).

How much doubt, Gold does not say, but the id/ego/super-ego model is hardly simple and it is only one of three models that Freud used to interpret personality structure. Moreover, Freud's model of psychological development is based on the ways in which children learn about, and learn to cope

---

1 Downs and Stea seem completely unaware of the personal and intellectual antagonisms between Freud and Jung, otherwise they would not conflate Freudian psychoanalysis and Jungian psychology. Nevertheless, Freud did suggest in his theories about the development of civilisation that archaic factors could still be detected in the mental life of children (see Chapter 6). This could be argued to constitute a 'racial memory', but it is not biologically transmitted in the way that Downs and Stea imply.

with, their world; and notions of mental illness and distortion are not psychoanalytic, but grounded in a discredited medical discourse. Gold, ultimately, replicates Downs and Stea's prejudices by accusing psychoanalysis of the view that nature was by and large unchanged by nurture.

The point for these critics of psychoanalysis was to guarantee the importance of intentionality, a modicum of free will and the capacity for self-awareness. So, I find it both ironic and paradoxical that an author whose work was widely used to underpin both behavioural geography and cognitive mapping should reach for Freud at the moment when he finds it necessary not only to distinguish between rat and man (sic), but also to guarantee men intentionality, a modicum of free will and the capacity for self-awareness. It is time to return to Tolman's work on cognitive maps. In Chapter 2, the story was abandoned at the point where Tolman suggested a level of explanation which stimulus-response theories could not account for. He suggested that 'the devils underlying many of our individual and social maladjustments can be interpreted as narrowings of our cognitive maps due to too strong motivations or to too intense frustration' (1948: 48).

From somewhere, intense feelings squeeze cognitive maps. The forcible contraction of cognitive maps produces devils which underlie individual and social maladjustment. Geographers seized on the idea of cognitive maps, but seem to have missed both the idea that there are 'too strong motivations' and 'intense frustration' and that these create devils which play havoc on the mind. Tolman suggests three modes of these over-strong motivations: 'regression', 'fixation' and 'displacement of aggression onto outsiders'. Tolman believes that a failure to grasp the reasons for these (dare I say) unconscious motivations could be disastrous:[2]

> We dare not let ourselves or others become so over-emotional, so hungry, so ill-clad, so over-motivated that only narrow strip-maps will be developed. All of us in Europe as well as in America, in the Orient as well as in the Occident, must be made calm enough and well-fed enough to be able to develop truly comprehensive maps, or, as Freud would have put it, to be able to learn to live according to the Reality Principle rather than according to the too narrow and too immediate Pleasure Principle.
>
> (Tolman, 1948: 50)

Well-behaved people, like well-behaved rats, are well-fed and therefore calm.[3] In this contented state, Tolman suggests, people can produce 'truly comprehensive maps': that is, when they are not deprived of food, or made to suffer in stressful conditions, people can learn to live by motivations other

2 There is a political desire here to produce new and better cognitive maps which is disquietingly reminiscent of Fredric Jameson's aesthetics of cognitive mapping (1991). I will follow up this point in the Conclusion to this book.

3 I hasten to add that this argument, and its moral overtones, is not obtained from Freud.

than the need to satisfy basic instincts. They can learn to adapt to reality, rather than slavishly responding to the need to satisfy animalistic pleasures. It is not surprising that the quick reference to Freud was missed by geographers, but Tolman's invocation of Freud does suggest a different understanding of psychoanalysis than the one presented so far.

In this account, the individual's mental development – significantly, revealed in the state of their mental maps – and their resultant behaviour are connected to two coupled principles of mental functioning, each of which modifies the other: where the Pleasure Principle relates to the search for satisfaction, and the Reality Principle refers to the circumstances found in the external world. Clearly, for Tolman, psychoanalysis explains the embeddedness of both mental functioning and overt behaviour simultaneously in the conditions of the external world and in the needs of the body. For Tolman, the mental map reflects the individual's embracing of the reality principle and the pleasure principle: psychoanalysis matters – it places people in a psycho-physical field, complete with mental maps! Intriguingly, Smith also picks up on this 'dialectic' between reality and pleasure – or between natural instincts and cultural prohibitions – in his account of Freudian psychoanalysis.[4]

In his book on the relationship between the city and personality, Smith discusses Freud's work on intrapsychic processes and his analysis of civilisation and individual psychology.[5] At first sight, it seems odd that Smith's concerns with urban life and planning theory should necessitate a discussion of Freud: after all, Freud made no great statements about either urban life or state intervention. On the other hand, as Smith argues, Freud's work does carry with it examples of urban life and an urban imagination. This is not Smith's principal reason for engaging with Freudian theory though; instead, it is psychoanalytic discoveries concerning the sources and causes of alienation which are at issue. If alienation, unhappiness and distress are the result of intrapsychic processes, then there is no point in designing and building the utopic city. Freud's theories, for Smith, might help illuminate the sources and causes of human misery and, therefore, inform an appropriate political programme. Psychoanalysis can demonstrate 'the pervasiveness of repressive social forms, the possibilities for developing critical self-awareness, and the centrality of both work and play to human development' (M. P. Smith, 1980: 52).

The line of argument which Smith develops is connected to Freud's id/ego/super-ego topology. The story goes as follows. The individual has to

4 So far as I am aware, Smith does not cite Tolman – the coincidence is down to Freud.

5 Smith mainly uses four of Freud's texts, 'The unconscious' (1915a), 'The ego and the id' (1923), 'The future of an illusion' (1927) and 'Civilization and its discontents' (1930). Smith employs the first two papers in his account of Freud's understanding of intrapsychic process, while the last two are held to be representative of Freud's explanation of the relationship between civilisation, culture and the individual. The framework for Smith's interpretation of Freudian psychoanalysis seems to be largely derived from Marcuse's *Eros and Civilization* (1956).

cope with two related, but opposed, biological instincts: the sexual and the aggressive. These instincts are located in the id. Because people cannot survive if they spend all their time pursuing either their sexual or their aggressive impulses, civilisation imposes strict prohibitions and taboos on behaviour. Each individual develops a super-ego, which makes them feel guilty if they have 'bad' thoughts or do 'bad' things, and which forces them to repress their instinctual impulses. The super-ego is profoundly social because its content is derived from cultural norms and laws. There is a conflict between the desires and angers of the id and the prohibition and taboos of culture. This struggle is played out unconsciously, but the ego steps in to negotiate between the feelings provoked by this struggle. The individual is never an unwitting dupe either to their impulses or to social conditions, partly because the id and the super-ego are in dialectical tension, and partly because the sexual and aggressive instincts are in dialectical conflict too. From this perspective, the ego is fundamentally alienated not only from other parts of the mind but also from the external (urban) world. As a result, you cannot expect happiness, just common misery.

For Smith, this bleak assessment carries fundamentally reactionary political implications: just as the ego has to adapt to reality and forgo pleasure, because it cannot effect intrapsychic processes, so politics and planning has to side with civilisation and repression, although it should enable people to relieve any unnecessary distress. For him, there are several defects in Freud's assessment of intrapsychic processes and their relationship to everyday life. Smith argues, first, that Freud fixed the real determinants of life in biological impulses, such that history becomes a backdrop to the vicissitudes of instincts; second, that Freud tended to mystify the causes of human misery by locating them in the mind, rather than in the realities of social oppression and exploitation; third, that Freud based his account of the balance of instinctual impulses on classical economics; fourth, that Freud failed to realise that his model of the mind was based on dominant notions of the bourgeois individual. In the end, Smith suggests (in an echo of other geographers' assessments) that 'it is thus unnecessary to reach the unknown depths of the unconscious mind, when historical realities provide a direct, adequate, and demonstrable explanation' (M. P. Smith, 1980: 213).

The example Smith uses to demonstrate this is the disturbances in US cities in the 1960s. For him, it is demonstrable that the direct causes of the black urban rebellion are located within the conditions of social, political and economic inequalities that these people were undergoing. The explanation of urban, race and class violence and conflict can be adequately explained through descriptions of historical, social and geographical inequalities. It is of course impossible to explain urban black rebellion without an account of the inequalities of power and material circumstances: the question is whether this can *adequately* and *directly* explain the ways in which people act within, react to and change their circumstances – not all poor and/or black people become rebels. I would argue that it is also necessary (though not enough) to understand the mental processes and this can be demonstrated by looking back at Smith's four criticisms of Freud.

First and second, Smith seems to forget that intrapsychic processes are constituted not just by the vicissitudes of instincts, but also by social prohibition and taboos. The implication is that power is not simply external to the individual, but fundamentally implicated in the development of the psyche. Whether someone reacts to a situation with violence will not only depend on material deprivation, but on how they feel about it. If people's actions are to be understood, then it is necessary to understand that feelings, impulses and thoughts are somewhere in the flesh,[6] but that people act and react to their feelings, impulses and thoughts – both consciously and unconsciously. Moreover, as Smith also seems to forget, the misery of the discontents is located within civilisation (and within the bourgeois family), so distress cannot be purely intrapsychic.[7]

Third and fourth, Smith is right to point out that Freud's notions of the individual and the mind are embedded in their social historical and geographical circumstances, which Smith takes to be primarily the Viennese middle-class family.[8] Smith fails to observe that the three topologies of the mind (economic, id/ego/super-ego, and unconscious/preconscious/conscious) are used by Freud as metaphors (albeit supposedly scientific ones) for different aspects of the mind. These topographies map out three individuals and not one bourgeois one. Moreover, as Smith knows, the drives are in dialectical tension and this disrupts any possibility that Freud's psychology can secretly settle on a stable heterosexual competitive greedy bourgeois identity.

Smith's points cannot be ignored, but they relate to two separate/d versions of psychoanalysis: on the one hand, id psychology, which interprets behaviour through an appreciation of the secret lives of the sexual and aggressive instincts; and, on the other, ego psychology, whose therapeutic practice stresses the adaptation of the ego to the external world, but does not help people to change those circumstances. Smith shows that interpreting urban life through either id psychology or ego psychology will produce a one-sided account. For me, the problem is that if concepts like 'id' and 'ego' are segregated, they not only lose their dynamics, but they also become naturalised and reified as static 'things-in-themselves', rather than being possible relationships. Moreover, as I understand it, the topologies of the

---

6 It can also be observed that feelings are not always in the same place – why else are thoughts commonly experienced in the head, while feelings are in the heart and stomach? This is partly why I argue that psychoanalysis involves a spatial imagination – the body is, has and makes a space.

7 There is an interesting link here to Tolman's version of the Reality/Pleasure tale. For Smith, pleasures are a luxury, whereas for Tolman they are a basic instinct. For Smith, material deprivation forces people to exist according to reality, whereas for Tolman they act according to pleasure. Oddly, Tolman's version is closer to Freud's. From this perspective, pleasure only means the satisfaction of a need. But, if pleasure is taken to have other more significant meanings, then Smith's account becomes more telling.

8 As I will show, Freud's imagination is also implicated in other social relations of power.

mind are not just constituted through internal spatial relationships, but are also themselves implicated in intricate and dynamic spatial relationships both with each other and with external worlds (which are also spatial). These are the spaces of the subject. The idea that psychoanalysis can contribute an understanding of simultaneously real, imaginary and symbolic spatialities has recently begun to intrigue cultural geographers, such as David Sibley and Gillian Rose.

PSYCHOANALYSIS *FOR* GEOGRAPHERS

As Smith highlights the perils of producing a one-sided Freudian interpretation of the relationship between the subject, space and the social, he raises significant issues. One question concerns the extent to which psychoanalysis can aid an explanation of social inequality. The example which Smith uses is the marginalisation and rebellion of black people in US cities. In Britain Sibley has used psychoanalytic concepts to understand the marginalisation of gypsies. Sibley blends the work of object relations theorists into an analysis of the ways in which boundaries are formed between the self and the other, the reasons for boundary formation and the subsequent marginalisation of specific groups as 'outsiders'.

Sibley neatly introduces the parameters of the problem this way:

> Dirt, as Mary Douglas (1966) has noted, is matter out of place. Similarly, the boundaries of society are continually redrawn to distinguish between those who belong and those who, because of some perceived cultural difference, are deemed to be out of place. The analogy with dirt goes beyond this, however. In order to legitimate their exclusion, people who are defined as 'other', or residual, beyond the boundaries of the acceptable, are commonly represented as less than human. In the imagery of rejection, they merge with the non-human world.
>
> (Sibley, 1992: 107)

In Smith's terms, the landscape of exclusion, which racialises the inner city and produces black urban oppression, can be read adequately and directly out of the politics and economics of social inequality. In Sibley's account, there has to be some explanation of why certain people are seen as 'other', why they are marked as 'other' according to 'some perceived cultural difference', why they are feared and loathed and why, consequently, they are seen to be 'out of place' and held at a distance (through the social regulation of space). There are two interrelated ways in which Sibley works through these questions: first, through a theoretical discussion of object relations psychoanalysis; and, second, through an analysis of the social construction of boundaries, which mark inside and outside, in the contexts of gypsy communities and childhood (see Sibley, 1981, 1995a).

In his work, Sibley describes a situation where gypsy communities are seen as 'deviant' by 'normal' society. Gypsies are constructed as dirty, dangerous and disorderly by dominant society. Wherever gypsies turn up, they are always seen as 'out of place' and conflicts arise to move them on, to purify space. This conflict is not just about the power relations between two communities, it is also about the control of space, but – even more – it is felt as a mutual violation of the body and spaces. These conflicts are as much internal as external: the racism which surrounds the arrival of gypsies is deeply felt, as pollution. This sword cuts both ways: 'While Gypsies are seen as polluting spaces controlled by dominant society, gauje [non-Gypsy] practices pollute Gypsy space' (1992: 112).

While gypsies are in a less powerful position, the psychodynamics of violation works across the spaces of power and exclusion. Sibley suggests that this marking of specific people as 'pollution' is involved with the simultaneously symbolic and social organisation of space. It works this way: gypsies have different beliefs and different spatial practices which seemingly threaten the social order (physically, metaphorically and mythically) in travelling across borders, in evading regulations and in refusing to settle into accepted patterns of home and work. In Britain, gypsies become matter out of place – and, because they are seen as 'other' and 'out of place', it becomes legitimate for the authorities to regulate, control and police the spaces where gypsy families are permitted to stay, live and work. Through eviction, gypsy communities are forced to live within this regime, but they resist and subvert this control by making even these spaces conform to their beliefs. Nevertheless, the regulation of space limits the opportunities for gypsies, discourages interactions between gypsies and gaujes, and restricts their family and social life.

For Sibley, 'space is an integral part of the outsider problem. The way in which space is organised affects the perception of the "other", either as foreign and threatening or as simply different' (1992: 116). The construction, maintenance and policing of spatial boundaries is not just a question of political economy, it relates to the ways in which people develop boundaries between self and other (following Keith, 1987 and S. Smith, 1990): these relationships are spatial, and they are also learned through childhood experiences (Sibley, 1995a). Sibley argues that larger societies' reactions to gypsies are based on the identification of gypsies as 'other', so a question which must be addressed is how people learn to recognise the differences, or boundaries, between self and other. It is at this point that Sibley draws on object relations psychoanalysis, particularly the work of Julia Kristeva (1980), Constance Perin (1988) and Melanie Klein (1959).[9] The important link which Sibley makes is between the body, the constitutive ambivalence of self, and the construction of (deeply-felt) emotional geographies. For Sibley, 'object relations theory provides us with a map of the self in place, an integration of the spaces of the body, the space of the self and the other, and

9 He also draws on the work of George Herbert Mead, Erik Erikson and D. W. Winnicott.

the mediating environments of the home, the locality and the world beyond' (1995a: 125).

In object relations theory, the child gradually develops a sense of its self, initially as a result of its gradual realisation that it has a separate body from the mother. The child's self is constructed as a relationship between self and other objects, such that the boundaries between self and the external world become increasingly well demarcated, but where the child's sense of self (or ego) is always mediated through its relationship to objects. The first point to note here is that this relationship is already social and the second point is that the word object refers to all objects – including carers, fantasies, wishes, bodily sensations, toys and so on.

Once the child realises that it is not connected to the mother's body, it dawns on the infant how helpless it is. It is argued that the child's experiences of loss, separation, abandonment, insecurity, desire, fear of dissolution and anger produce an ambivalent response to the mother. On the one hand, the mother provides food, comfort, love and warmth – and she is a 'good' object. On the other hand, she leaves the child on its own, disappears from sight, and does not instantly provide gratification when it is wanted – so she is also a 'bad' object. The mother is loved for the love she provides *and* she is hated for the love she withdraws. The child, however, cannot comfortably hate, or be angry with, the mother, because of the fear that the mother may punish it. Since neither emotion can be unconditionally expressed, the child is forced to control both its love for, and hatred of, the mother. Nevertheless, these ambivalences rattle around the child, structuring its responses to senses of loss, abandonment, comfort and love later in life: 'Aversion and desire, repulsion and attraction, play against each other in defining the border which gives the self identity and, importantly, these opposed feelings are transferred to others during childhood' (1995a: 125).

Children learn to displace dangerous or unwanted feelings onto others – others who are perceived to be different. These feelings are simultaneously social, bodily and spatial.[10] The concept which Sibley uses to describe this relation is abjection (from Kristeva, 1980). In abjection, senses of revulsion over bodily materials and feelings are established. The subject wants to expel whatever is reviled, but is powerless to achieve it: thus, for example, the desire to be completely clean all the time cannot be achieved, purity cannot be maintained. Abjection, then, is a perpetual condition of surveillance, maintenance and policing of impossible 'cleanliness'.

Abjection brings several elements together: first, it aligns the subject and space, the biological and the social (see Perin, 1988); second, it is learned: whatever is considered to be pollution or taboo varies according to cultural and social values (see Douglas, 1966).[11] When people are confronted by groups who are valued by wider society as 'dirty', then they are capable of

---

10 This point resonates with arguments made in Chapter 6 below.
11 I will pick up this discussion in Part III of this book.

reacting with embodied rage and hatred, venting seemingly legitimate emotions, which have been 'under wraps' since childhood. It is in this context that racist Nazi propaganda which drew associations between Jews and rats can be understood – and, in the context of white images of black slaves in southern US states, that black urban rebellion might be reinterpreted.[12]

Sibley builds an intricate and dynamic picture of the relationship between intrapsychic processes, the social organisation of space and power relationships. More than this, he also specifies the 'dialectic' between the individual and the social – which proved so illusive in Part I above – through the Kleinian concepts of 'introjection' and 'projection'. The two-way street between the individual and the social has, in this account, two sides:

> introjection and projection function from the beginning of post-natal life as some of the earliest activities of the ego, which in my [Klein's] view operates from birth onwards. Considered from this angle, introjection means that the outer world, its impact, the situations the infant lives through, and the objects he encounters, are not only experienced as external but are taken into the self and become part of his inner life ... Projection, which goes on simultaneously, implies that there is a capacity in the child to attribute to other people around him feelings of various kinds, predominantly love and hate ... Thus an inner world is built up which is partly a reflection of the external one. That is to say, the double process of introjection and projection contributes to the interaction between internal and external factors. This interaction continues throughout every stage of life.
>
> (Klein, 1959: 250)

While inner life is distinct, there is continuous exchange between the internal and the external, but this 'dialectic' is itself interacting with the transactions between 'introjection' and 'projection'. The inside and the outside, the spaces of the ego and body, the adult's understanding of its own feelings, and people's responses and reactions in the world are enmeshed with intrapsychic processes, social values and power relations. Sibley's understanding of the landscape of exclusion shows not only how wide of the mark many criticisms of psychoanalysis have been, but also just how limited is a purely political economy of, for example, the inner city. While Sibley's work has far-reaching consequences, the 'landscape of exclusion' can also be interpreted using other psychoanalytic tenets. Initially, like Sibley, Gillian Rose drew on Kristeva's account of abjection to understand the ambivalence towards the (female) body in dominant geographical discourses (G. Rose, 1991). However, Rose has developed her radical and subversive critique of

12 This account ties in with Klaus Theweleit's discussion of Nazi male fantasies, see Chapter 6 below.

masculinist geography through an engagement with psychoanalytic feminisms.

Rose's involvement with psychoanalysis is, like Sibley's, strategic and used to interrogate and subvert practices of power and exclusion. The specific 'object' of her suspicions is 'the Same/Other structure of masculinist geographical discourse' (G. Rose, 1993: 62). Using a mixture of Chodorow's and Irigaray's account of mothering,[13] she suggests that boys and girls are placed differently into positions of knowledge and power and, further, that these positions imply different relationships to space and the body (see 1993, Chapter 4).[14] Rose emphasises the importance for boys of gaining autonomy, of their fear of castration and of their need to gain distance in order to protect them from their desires. Moreover, the identity of boys is based not only in an ambivalent structure of fear and desire, but also in the need to gain distance from, and control over, objects in the world. Furthermore, boys' relationships to their mothers and fathers forces them to take up a position in a Same/Other economy of meaning and power. This story parallels that told by Sibley. For Rose, significantly, this is also a sorry tale of sexual differentiation (in the same way that it marked other kinds of difference for Sibley). Accordingly, boundaries between Same and Other are instituted for boys which are not the same for girls – mainly because girls do not have to separate themselves from their mothers, and they can develop themselves in relation to others in ways which boys have to refuse. For her, masculine and feminine positions are different: different spaces, different imaginaries.

The hierarchical division of the world into the Same and the Other, Rose suggests, masks a radical uncertainty about identity. For her, Lacan's reading of Freud foregrounds this radical uncertainty and adds a sense of the impossibility of understanding identity that was somewhat marginalised in Chodorow's recovery of mothering (G. Rose, 1993: 77–78). In Lacanian psychoanalysis, subjectivity is installed as a cover for the fear of loss, initially of the mother. The mother's body becomes a significant site/sight for the child. When it is seen that the mother lacks a penis, this confirms the reality of castration. It is only the visual which provides information about the mother's already-castrated state. For the boy, the look provides essential evidence about the identity and the constitution of the object (as an object of desire or disgust). The child thinks that the eye provides all the information about the body that it needs, but this is an illusion. The boy's identity is trapped in a radically unstable regime of looking: unstable because it seems to provide a complete knowledge, but is actually partial. From this perspective, masculine knowledges are locked into an Imaginary world that is based on a morphological uncertainty about the Same and the Other.

Rose's work comprises a radical re-imag(in)ing of the morphology of

---

13 See Chodorow, 1978 and Irigaray, 1977.

14 In her later work, Rose has suggested that the language of 'position' itself is implicated in a masculine imaginary of 'mappable' space (see G. Rose, 1995).

geography, using many sources, but I will pick out the way she develops this argument psychoanalytically: first, through a critique of masculinist geographical discourse; and, second, through opening up the possibilities of enacting new feminine spaces. For Rose, psychoanalytic feminism shows how the map of subjectivity, language and power is drawn such that masculine positions are characteristically dualistic and ambivalent. She describes the ways in which cultural geographers, including Cosgrove and Daniels (see pp. 59–61 above) have come to see the landscape as just like the female body: their work is saturated with gendered and (hetero)sexualised tropes. Rose shows that this sexualised looking is neither easy nor without contradictions. Indeed, she demonstrates that there is a constitutive ambivalence of the masculine gaze. On the one hand, there is a fear of the Other (like the 'castrated' mother), which can express itself as a desire for critical distance on the object studies or the requirement for objective knowledge. On the other hand, there is a desire to get close to the object of study, to know it, to get under its skin as it were. Masculinist geographical discourse oscillates (violently) between these positions, never able to tear itself away from the desire to look, to do fieldwork, 'to get a bit of the other'.

Mother Nature and Mother Earth are implicated in this ambivalence. Symbolising the landscape as feminine produces a desire to see, to watch and assert masculine sexual control, but landscape also dangerously seduces, refuses to lie down and behave as it should. Mother Earth cannot be both Mother and Mistress (following Kolodny, 1973).[15] The feminisation and (hetero)sexualisation of objects means that looking is characterised by a sexual regime of the visual. This argument is supported by Sibley in terms of other kinds of difference. He argues that stereotypes of gypsy communities are predominantly visual: usually placing gypsies in bell-topped, horse-drawn carriages in rural settings (Sibley, 1992). For Sibley and Rose, dominant practices of knowledge and seeing produce pictures of 'the real' which say more about the fantasies of the observer than the 'realities' of the observed. The terms 'real' and 'realities' are in scare quotes because regimes of looking only seem to produce objective knowledge about the world; a fantasy which is characteristically, for Rose, masculine. In order to interpret these masculinist regimes of the visual, she turns to post-Lacanian psychoanalytic feminism. Rose argues that 'the desire for full knowledge is indicated by transparency, visibility and perception. Seeing and knowing are often conflated' (1993: 86).

The gaze is eroticised by a desire which is at once expressed and repressed. The masculinist gaze has two, related pleasures: narcissistic and voyeuristic (following Mulvey, 1975). In narcissism, the viewer sees the image in relationship to himself; while, in voyeurism, the observer holds the object in view but at a distance. In both situations, the look is active and the object is

---

15 Rose also cites the work of Klaus Theweleit (1977): his discussion of 'earth' is described in Chapter 6, p. 204.

passive. Within the dominant sexual politics of looking, the active look is encoded as masculine and the passive object is feminised. Women appear, men look. From this perspective, 'geography's pleasure in the landscape can be interpreted through the psychoanalytic terms across which the gaze is made – loss, lack, desire and sexual difference' (1993: 104).

Where geography professes a desire to look at the world, it secretly harbours not only a hierarchical dichotomy between the Same and the Other, but also a set of allied oppositions between the masculine and the feminine, between the active and the passive, and between the real and the metaphorical. Rose's project is to undo, subvert and transcend these power-infused dualisms.

Strategically, she has begun to think through the relationship between the feminised object, the discursive production of objects and the possibility of producing other knowledges. Rose suggests that the apparent coherence of sexual identities is founded on the repetition of themselves as coherent and identical (following Butler, 1990, 1993). She refuses to play these monotonous, repetitive boys' games. She refuses a position where women are identified as nature, where the earth is seen as mother. For her, Irigaray suggests other spatialities for the feminine subject. These are to be found in different voices, in shifts of place. She quotes Irigaray: 'The transition to a new age in turn necessitates a new perception and a new conception of *space–time*, the *inhabiting of places*, and of *containers*, or *envelopes of identity*' (Irigaray, 1993: 7; cited in G. Rose, 1995).

The possibility of producing radical and subversive knowledges, which do not rely on the significance of the phallus, masculinist regimes of looking and the psychodynamics of sexual and gender oppression, requires a new Imaginary, finding new voices and inhabiting new places. In Rose's writings, psychoanalysis informs the articulation of at least one of these voices, one of these places.

SPACES OF THE SUBJECT

The work of Sibley and Rose provides a politically-embedded flow of psychoanalytic ideas for thinking about the relationship between subjectivity, society and space. There are hints, here, of a 'psychoanalysis of space'. Their use of different aspects of the mother–child relationship, their understanding of the psychodynamics of social situations and their sensitivity to power relations are both inspiring and insightful. In particular, I think that Sibley shows how personal boundaries and social boundaries not only carry deep emotional resonances but also play out through geographical space. Meanwhile, Rose shows how knowledge about ourselves and about the world is constituted in the ambivalences of intrapsychic processes and how this configuration relates to some but not others. Intriguingly, she toys with the trope of the mirror to show how space is marked as a surface, a depth and a wall, but nevertheless how it could become fluid, bleed and be something else. There are also shared questions of identity too: for Sibley, identity is formed through the individual's constellation of object relationships, while for Rose, it is a performance, a masquerade. In each, there is no sense that identity relates to some kind of preformed essence. The subject is both social and spatial. Space, in their view, is never 'real or metaphorical', but both real and metaphorical and more. Moreover, both Sibley and Rose show that there is an intimate relationship between the body, regimes of the visual, and the social organisation of space. In various ways, I will pick up on these ideas throughout the rest of this book.

Very often in psychoanalytic discussions, there is resort either to Freud or Lacan, or to narratives about the child. In this part of the book, I explore the work of Freud and Lacan and their respective accounts of childhood. My versions of Freudian and Lacanian psychoanalysis are not innocent, however: I read them both in order to disclose insights into the different kinds of spatialities through which subjects become subjects. These are not literature reviews, nor are they faithful repetitions of the master's voice. The reading is doubled up, moreover, by the reading of different spatialities *into* Freudian and Lacanian interpretations of the relationship between identity, meaning and power.

In my reading of Freud and Lacan, I have tried to produce an interpretation which is open to the fluidity and fixity of subjectivity – and, beyond this, spatiality. The problem for me has been that Freud and Lacan normally assume that the child is a boy. I have chosen to relate their narratives of child development as if the child were an 'it'. My intention here is to allow the child to become different, to become any of a thousand tiny sexes (to use Grosz's expression). Ironically, the downside of this strategy is that it displaces Difference – the 'it' never becomes *a* boy, or *a* girl, or . . . and so on. Nevertheless, this tactic enables difference to be seen as constitutive of subjectivity, without giving *a priori* significance to any particular kind of difference. These difficulties are encountered throughout this book.

Part II, 'Spaces of the subject', has two chapters and a conclusion. The first chapter looks at the ways in which Freud placed the individual within specific mythologies, which partly drew on and partly informed his conclusions about intrapsychic processes and the circumstances with which children had to cope. The second chapter assesses the ways in which Lacan located the 'I' in the mistaken place both of the specific other and the generalised other. Conclusions are drawn not only about the psychoanalytic spaces of the subject, but also about the ways in which these might inform a psychoanalysis of space. In order to specify these relationships further, the work of Freud, Lacan and Lefebvre is triangulated. This map enables me to chart six points of reference for an interpretation of the psychodynamics of place.

4

MYTH PLACED
the traumas and dramas of childhood

In his article 'Geography, experience and imagination', Lowenthal (1961: 249) extracts this passage from Aldous Huxley's *Heaven and Hell*:

> our mind still has its darkest Africas, its unmapped Borneos and Amazonian basins ... A man consists of ... an Old World of personal consciousness and, beyond a dividing sea, a series of New Worlds – the not too distant Virginias and Carolinas of the personal subconscious ... the Far West of the collective unconscious, with its flora of symbols, its tribes of aboriginal archetypes; and, across another, vaster ocean, at the antipodes of everyday consciousness, the world of Visionary Experience ... Some people never consciously discover their antipodes. Others make an occasional landing.
>
> (Huxley, 1956: 69–70)

For Lowenthal, this passage marks the necessary intersection between the unknown territories of the mind and a specifically geographical understanding of the mind. More can be gleaned from this allegory; it can, for example, be read in a way that striates the work of Sigmund Freud. Firstly, however, it ought to be noted that, at the time of writing, Huxley was living in California and experimenting with 'mind-expanding' drugs such as mescalin and LSD, which he asked for on his death-bed. Nevertheless, Huxley's use of geographical metaphors is more forcible than Lowenthal's clipping reveals. A slightly different editing reads:

> Like the earth of a hundred years ago, our mind still has its darkest Africas, its unmapped Borneos and Amazonian basins ... It is difficult, it is all but impossible, to speak of mental events except in similes drawn from the more familiar universe of material things. *If I have made use of geographical metaphors and zoological metaphors, it is not wantonly, out of a mere addiction to picturesque language. It is because such metaphors express very forcibly the essential otherness of the mind's far continents, the complete autonomy and self-sufficiency of their inhabitants.* A man consists of what I may call an Old World of personal consciousness and, beyond a dividing sea, a series of New Worlds – the not too distant Virginias and Carolinas of the personal subconscious ... the Far West of the collective unconscious ... and,

> across another, vaster ocean, at the antipodes of everyday consciousness, the world of Visionary Experience ... Some people never consciously discover their antipodes. Others make an occasional landing.
>
> (Huxley, 1956: 69–70, emphasis added)

He ends this passage by saying that there are two basic ways of exploring the '*terrae incognitae*' of the mind (in Huxley's words, page 71); first, by the use of hallucinogenic drugs; and, second, by hypnosis. Huxley is aware that neither route is perfect (page 70). Although Freud's researches into the '*terrae incognitae*' of the mind began in an earlier century, he too experimented with both drugs (cocaine) and hypnosis and, like Huxley, had found them less than perfect for providing routes into the mind. Perhaps Huxley's navigational methods for exploring the mind were a response to the failings of psychoanalysis, but Freud's dynamic psychology exists as a method because of his desire to go beyond the limitations of a reliance, like Huxley's, on quick fixes.

As a way of introducing Freud, we can start with Huxley's map of the mind, partly because they share a predilection for understanding the mind through geographical (or spatial) metaphors: hence the resonance between Huxley's imagery and Freud's (in)famous use of the 'dark continent' metaphor to describe femininity (see Doane, 1991; see Chapter 6 below). By way of warning, we should note that their use of these metaphors grounds their ideas of the mind in patriarchal and imperialist presumptions about space and place – therefore this chapter must read Freud against the grain, if it is to set out a position from which to 'explore' the body and the city.

Huxley is evoking a kind of world order of the mind; naturalised, ahistorical, fixed, waiting to be discovered (should we dare), with nearby realms of consciousness and far off continents of the collective unconscious. It is not hard to think of Freud drawing on just such a territorial imagination to understand the dynamics of the psyche, from the too familiar lands of the conscious, to the unchartable territories of the unconscious. Freud was, after all, forever drawing on contemporary accounts by anthropologists to support his assertions about the so-called primitive mind, which he likened to the child's mind and to femininity through spatial metaphors. Moreover, this geographical imaginary must be located within other horizons: bourgeois, empire(s), male, masculine, Jewish, Christian, Vienna, the West. Freud's analyses of 'civilisation' are, then, circumscribed by his own hinterland of meaning, identity and power. Nevertheless his ideas offer ways of reconceptualising the relationship between the individual, others and the external world.

I will explore these ideas in the first instance by looking closely at one key article – 'Group psychology and the analysis of the ego' (1921). This article establishes a framework through which it is possible to develop the links between key Freudian ideas and a psychoanalytically-informed account of subjectivity, space and the social. The significant point, for me, is that whenever Freud discusses the bonds between individuals and groups, within groups and between groups and others, he demonstrates that the binds that tie are constituted simultaneously psychically, socially and spatially – and, because of this, these relationships play out differently in different situations. I have also chosen to detail this paper both because of where it stands in relation to Freud's evolving system of ideas and also because it seems to say so much about the rise of fascism, whether in Germany in the 1930s (aspects of which will be discussed in Chapter 6 below), or in the former Yugoslavia in the 1990s, or elsewhere.

Freud's analysis of the individual and the group stands at the gateway to his later ideas on the constitution of human drives and the fragmented composition of the psyche. Behind it lies an older psychoanalysis which is much more about the interpretation of meaning and the individual's relationships with other people (epitomised by Freud, 1900 and 1901). 'Group psychology and the analysis of the ego' can be read in two different ways: first, as a way of showing how civilisation distorts the expression of universal human desires (an analysis Freud himself pursues later; see Freud, 1930); and, second, as being about the ways in which people relate to each other and to authority, within dense networks of meaning and power (which Freud had examined in 1913). For me, Freud's work is most interesting where he discusses 'object relationships',[1] because it is here that the spaces of the subject are sketched and elaborated: of particular significance is his discussion of the modalities of identification.

Choosing to read 'Group psychology and the analysis of the ego' is slightly eccentric because it is 'Civilization and its discontents' (1930) which has had the more profound effect on critical social theory.[2] However, Freud's analysis of the relationship between the psychologies of the individual and the group suggests a number of psychic mechanisms which can be used to theorise the subject's location within the constellations of their simultaneously personal, social and spatial relationships. These mechanisms describe the ways in which individuals form their object relations with the external world and the ways the external world sanctions and/or prohibits particular object relations. So, 'Group psychology and the analysis of the ego' offers the possibility of a reading of the spaces of subjectivity which privileges neither the individual nor the social and which collapses neither into the other. Furthermore, the article presents a Janus-faced set of ideas from both Freud's early suspicions about meaning (revealed in dream analysis and in his use of

---

1 It is worth remembering that, in psychoanalysis, an 'object relationship' does not just refer to a relationship to a thing, but also to a person, an idea, fantasy, a part of the body, and so on.

2 In his analysis of civilisation, Freud is much more concerned with the civilising impulse as an innate feature of humanity. This has produced a separate set of ideas relating to the development and distortion of the instinctual life of individuals – this version particularly appealed to the Frankfurt School because it provided a standard against which freedom could be measured (see Held, 1980). See particularly Marcuse, 1956, 1964, but also Brown, 1959.

myths) and his later speculations on instincts (as seen in the counter-posing of Life and Death drives as constitutive of the psyche).

The juxtaposition of geographical and psychoanalytic imaginations is neither arbitrary nor bizarre. Freud used spatial metaphors in order to convey a sense of the materiality of the mind and of his understanding of it.[3] Thus, in reply to a fictional question asked by an imaginary 'Impartial Person' about the construction of mental apparatus, Freud argues,

> it will soon be clear what the mental apparatus is: but I must beg you not to ask what material it is constructed of. That is not a subject of psychological interest ... We shall leave entirely on one side the *material* line of approach, but not so the *spatial* one. For we picture the unknown apparatus which serves the activities of the mind as being really like an instrument constructed of several parts (which we speak of as 'agencies'), each of which performs a particular function and which have a fixed spatial relation to one another: it being understood by that spatial relation – 'in front of' and 'behind', 'superficial' and 'deep' – we merely mean in the first instance a representation of the regular succession of the functions. Have I made myself clear?
>
> (Freud, 1926a: 103–104)

Freud has the Impartial Person reply, 'scarcely', and then object to this 'strange anatomy of the soul'. Nevertheless, Freud insists that there is a relationship between the biological, the spatial and the mind, but where the mind consists of a number of 'agencies' (as we will see): partly, such a conception prevents the simple reduction of mental processes either to biology or to 'civilisation'. It can be noted that this rejection of mechanistic interpretations of the mind usefully parallels geographical critiques of fixed, undialectical and passive conceptions of space (see also Chapter 8 below).

Freud's pronouncements on the social have not inspired most geographers up to now, nevertheless it is conceivable that certain psychoanalytic propositions could be transplanted into a geographical imagination. The readings I provide are not intended to summarise Freud's work, but instead to identify elements in that work which might instigate a psychoanalytically-informed geographical imagination.[4] I am looking for two different things: first, a set of ideas that bleed dialectical notions of the individual and the social; and, second, a spatial imagination that helps say something about the world which has been difficult to express without psychoanalysis. Geographers have long known that space is socially constructed and that the social is spatially constructed (see Massey, 1994). It is known that space is not fixed and passive, but fluid and active (see Lefebvre, 1974 and Soja, 1989). Yet, the thinking through the spatialities of everyday life has been difficult to

3 For an analysis of Freud's spatial metaphors, see Brennan, 1992, especially pages 17–25.

4 There are many studies of Freud's work as a whole, which are generally available (see Ricoeur, 1970; Frosh, 1987; Wollheim, 1991; and Brennan, 1992).

articulate – perhaps because there has not been a language sufficiently adept at capturing the complex psychodynamics of place.

What follows, then, is an attempt to read the spatialities of psychoanalysis simultaneously out of, and back into, Freud's analysis of the individual, the group and the external world. The key point is that each of Freud's descriptions of group interactions involves distinct, specific spaces of the subject, where such spaces are characterised both by the ambivalences of identification and desire and by their embeddedness in social prohibitions, requirements and sanctions.

FREUD AND GROUP PSYCHOLOGY

In the introduction to 'Group psychology and the analysis of the ego', Freud makes it clear that there is a relationship between the psychology of the individual and the social world. Although Freud psychologises the social – that is, he assumes throughout that the social has a psychology – he refuses to collapse the individual into the social or to see the individual as the sum of her or his object relations. Nevertheless, individuals have to find a place in their object relationships:

> only rarely and under certain exceptional conditions is individual psychology in a position to disregard the relations of this individual to others. In the individual's mutual life someone else is invariably involved, as a model, as an object, as a helper, as an opponent; and so from the very first individual psychology, in this extended but entirely justifiable sense of the words, is at the same time social psychology as well.
>
> (Freud, 1921: 95)

Freud places individuals within the field of their relationships with others; importantly psychology is 'social' and 'spatial' in this sense from the very first. The new-born baby will unavoidably encounter its mother, followed by a series of other others such as the father, siblings, doctors, carers, and so on. However, Freud believes that the child will have a relationship to itself, which lies outside of these encounters with others. The next step in this analysis is to draw a distinction between social and narcissistic mental acts, though both of these occur within the mind of the individual. It is at this point that Freud introduces the issue of group psychology and what this might be. For him, group psychology is not concerned with the individual's relations to (significant) other people, but 'with the individual man as a member of a race, of a nation, of a caste, of a profession, of an institution, or as a component part of a crowd of people who have been organised into a group at some particular time for some definite purpose' (1921: 96).

Having defined the problem of group psychology and distinguished it from the question of the individual, Freud is then in a position to inquire into the quality of the 'group mind'. He begins this task with a summary of recent ideas, starting with Le Bon's description of the collective mind.[5] He argues that Le Bon fails to account for the reasons why individuals get involved with groups at all. So, Freud wants to know, why do people bind themselves into collectivities so completely and so deeply. In this, he agrees with Le Bon that individuals obliterate their conscious selves and give way to a 'racial unconscious', such that individual differences are submerged by group similarities. Characteristic of the group, they both believe, is a sense of invincibility, the immediacy of instinctual desires (both in terms of their expression and the need to have them satiated), anonymity, irresponsibility and a lack of restraint.

> In a group the individual is brought under conditions which allow him to throw off the repressions of his unconscious instinctual impulses. The apparently new characteristics which he then displays are in fact the manifestations of this unconscious, in which all that is evil in the human mind is contained as a predisposition.
>
> (Freud, 1921: 101)

For Le Bon, moreover, groups are contagious and hypnotic, such that individuals lose their 'civilisation': the individual becomes a barbarian in a crowd. Freud is clearly sympathetic to Le Bon's description of the crowd and the individual's mob behaviour: he repeats, apparently approvingly, the idea that mobs cannot tolerate delay between desire and fulfilment, are as intolerant as they are obedient, respect force (and not kindness), demand strong heroes, think in images, are open to suggestion and uncritical, and go directly to extremes. In this respect, both Le Bon and Freud make an explicit association between the mob and 'uncivilised' behaviour, which thereby makes the mob child-like and primitive since it allegedly lacks the requisite repression of unconscious desires that characterise civilised behaviour. Freud puts it this way: 'when individuals come together in a group all their individual intentions fall away and all the cruel, brutal and destructive instincts, which lie dormant in individuals as relics of a primitive epoch, are stirred up to find gratification' (1921: 106).

Further, by stressing the simultaneity of desire and fulfilment and the readiness of the group to believe illusions, Freud feels able to describe group psychology as neurotic, because neuroses are grounded in illusions which are in turn premised on unfulfilled wishes. There is here a shared map of meaning and power which denigrates the mob, children, neurotics and so-called primitives, though this is eventually worked through in different ways by Le Bon and Freud.

Freud agrees with Le Bon's suggestion that, in the mob, there is a need to obey, a need to believe in something and someone, a special 'hypnotic', 'charismatic' and 'fanatical' someone who provides strong leadership and beliefs. However, Freud is not content with this analysis of crowd behaviour

---

5 See G. Le Bon, 1895, *The Crowd: A Study of the Popular Mind* (translated in 1920, London).

because it fails to account for the role of the leader. Now Freud must not only explain the relationship between the individual and the group, but also correlate this explanation with the ties between the group member, the group and a specific individual: the leader. This correlation enables Freud to make a more wide-ranging critique of collective psychology.

Le Bon and others, Freud argues, only talk about short-lived groups – such as the mob – but ignore longer-lived groups, such as civil institutions: the crowd is, in this view, just 'a high but choppy sea to a ground swell' of society (Freud, 1921: 112). The missing leader has become central to the story of why groups form, why organisations are developed and why people submit to the rules and standards of groups. Although, by 1921, Freud had started into the third (and last) phase of his development of his theory of drives – in which he posited the existence of a death drive (see also Freud, 1920) – his interpretation of this problem was still based on an assessment of the nature of the individual's object relationships, such that the individual identifies with the group's submission to the leader. It can be noted that these object relationships are spatially constituted: thus, the leader is raised 'above' the crowd.

For Freud, the crux of the matter still lies in accounting for the formation of group organisations: the problem is that the group must be given the characteristics of the isolated individual, such that individuals feel able to lose themselves in the group. The question is why would an individual wish to abandon their sense of self-continuity, self-consciousness, independence and autonomy. The answer must lie in the necessity for a particular kind of attachment that the individual feels for the group and for the leader – a feeling that is so strong that it overrides any desire for individuality. For Freud, this involves an analysis of the economics of the libido: in a contemporaneous account ('Beyond the pleasure principle', 1920), 'libido' is the amount of emotional energy of the sex drives, the nucleus of which is sexual energy (Eros).

In other words, there is a 'love' between the individual, the group and the leader which both explains and needs to be explained in order to account for the strings that knot the group together. It should be noted that 'love' is neither simple, singular nor fixed, but complex, multiple and dynamic. It is a mistake to think that Freud is trying to account for groups solely through a 'sex drive' understood in crude terms. Instead, he is arguing that emotional ties bind the group and that these ties have the strength of love, and it is because of the strength of their feelings that individuals want to identify with groups and it is because they 'express' those feelings that groups seem to provide individuals with 'freedom'. This remains true even for groups which apparently lack any freedom or love for the individual, such as the church or the army.

It is the quality of love which also marks the mob as being childlike, neurotic and primitive, for it is in these states that individuals are free(d) of the ties of civilised behaviour: that is, the mob's wishes are liberated from the repressions of propriety. Meanwhile, groups such as the church or army are held together by an authority – God, the general – which loves, or at least treats, everyone equally, but without which they would dissolve. The ability of religion and the military to bind people together is founded on the illusion of love, demonstrated by the laws within the group. Freud argues that these hierarchical organisations set up a series of father–son structures: thus, in the army, the superior officer is 'a father who loves all soldiers equally, and for that reason they are comrades among themselves' (1921: 123). The strength of these internal ties within the group requires, on the other side of the coin, an intolerance of those outside the group.

Groups then are (spatially) defined by a constitutive inside and a constitutive outside, where the separateness and distinctiveness of the group can be assessed by the quality of its internal 'love' and external 'hate'; it is this that makes the group more than just a collection of individuals. So important is the need to 'love' and 'hate' that 'the narcissism of minor differences' becomes exceedingly important (1921: 130): 'of two neighbouring towns each is the other's most jealous rival; every little canton looks down upon the others with contempt' (1921: 131). One might detect such feelings in England between, say, Liverpool and Manchester or between, say, northerners and southerners, let alone the rivalries between the English and the Scots. There is a hierarchy of 'contempts' for the other and this hierarchy can be organised on the basis of geographical location and spatial scale.

With distressing insight, Freud claimed that psychoanalysis was 'no longer astonished that greater differences should lead to an almost insuperable repugnance, such as the Gallic people feel for the German, the Aryan for the Semite, and the white races for the coloured' (1921: 131). Later Freud is to note that similar attachments and disgusts can be detected between different classes, but he would not allow this analysis to be extended to gender relations, which he argued are constituted by the only true emotional ties. Groups are held together, then, by two kinds of emotional ties: first, those that are derived from sexual drives, which have been diverted from their proper aim, namely sex; and, second, those derived from death drives, which manifest themselves as hostility and ambivalence, though Freud feels unsure at this stage about the importance and source of death drives, arguing against himself that hostility may result from narcissism (i.e. self-love produces other-hatred).

Having suggested that a key mechanism in the formation of groups is the attachment (object-cathexis) of libido to the group and to the leader, Freud proceeds to offer another mechanism capable of tying people into groups: identification. Identification is 'the expression of an emotional tie with another person. It plays a part in the early history of the Oedipus complex' (1921: 134; see below). For Freud, the Oedipus complex is central to the formation of adult sexuality, but identification prepares the way for the child's entry into the Oedipus complex, whereby the boy actively identifies with the father and takes him as his ideal (see also Freud, 1914: 81–84). The form of identification learnt at this stage sets the pattern for the boy's later object relationships.

There are two distinct patterns of emotional ties: first, identification with the father; and, second, attachment to the mother. These exist side by side for a time, without contradiction, until the Oedipus complex (see below), where and when they are brought together. The boy sees the father as an obstacle to his desire for the mother and wishes to dispose of the father and replace him. Identification, then, is ambivalent: marking both tenderness and hostility. Freud's analogy involves 'the cannibal' who has a devouring affection for his enemies.[6] At this time, in the early 1920s, Freud still supposes that the (civilised) girl's development is parallel to the boy's: the girl identifies with the mother and is attached to the father. Therefore, identification, and its ambivalences, is universal – from primitive to bourgeois, from boy to girl.

Elsewhere in Freud's writings, the processes of identification are differentiated and allied to repression. Five key concepts are used: identification, incorporation, idealisation, introjection and internalisation, each marking a different mode through which individuals take on the characteristics of their chosen ideal. In identification, the subject assimilates an aspect of the object and this transforms the subject, wholly or partially, along the lines provided by the idealised ideal: personality is constituted and specified through identifications. In incorporation, the subject, in fantasy, has an object penetrate the boundaries of the body and keeps the object 'inside' the body (see Freud, 1905). In idealisation, the individual judges an object to be in a condition of perfection; identification with this idealised object contributes to the formation of ego-ideals, and the agency of the super-ego (see Freud, 1914, 1921). In introjection, the subject, in fantasy, takes objects and their qualities from the 'outside' and places them 'inside', but in contrast to incorporation this does not involve bodily boundaries (see Freud, 1915b). In internalisation, the subject transforms intersubjective relationships into 'internal' relationships; thus, external conflicts become conflicts within the individual. (Terms such as 'super-ego' and 'object' have specific meanings in psychoanalysis, so they will be defined below.)

These modes of identification have significant implications for a geographical imagination: first, because they are all spatial, where space is not a passive backdrop, but is constitutive of these relationships; second, because they describe specific relationships between the body, the mind and space; third, because they are all 'dialectical' in some sense; and, last, because they disrupt cosy assumptions about the 'nature' of interior space and exterior space. Through these modalities of identification, the individual institutes a personality along the lines of models provided by the external world. In these terms alone, it will be obvious that personality formation

---

6 The invocation of the figure of the 'cannibal' once more marks the foundations of his theories in its period, linking them to vituperative images of 'savage' and 'primitive' others as represented in imperial cartographies of meaning and power (see Chapters 6 and 8 below, and Fuss, 1995).

is hardly simple (as geographers have implied).

It would be better to say, instead, that personality is initiated, constituted and specified through conflicts, prohibitions, inconsistencies, ambivalences and incoherences, broken across many boundaries and striated by many lines of meaning, identity and power. In personality formation, it is also important to note the centrality of 'loss', where this must also be understood spatially (see the discussions of the Fort/Da game, pp. 134, 168 and 227–228). The child is perpetually losing objects of desire: first and most important is the loss of, and the renouncing of, the mother. In order to defend itself against these losses, the child introjects or incorporates the lost object. So, personality is founded on mourning and melancholia (see Freud, 1917b). Thus, for example, 'a child who was unhappy over the loss of a kitten declared straight out that now he himself was the kitten, and accordingly crawled around on all fours, would not eat at table, etc.' (1921: 138). The child can compensate for the loss by (potentially) permanently taking on the characteristics of the lost object; the lost object can then be set up as an ideal for the subject, which is constantly found lacking against the standard of the perfect ideal.

In this case, Freud presumes that the psyche is split into distinct 'agencies', each agency raging against the other, but where one agency contains the lost object: in Freudian terms, one agency, *the ego*, sets up another agency, *the super-ego*, which both contains the idealised lost object and acts as a moral overseer (see Freud, 1940). As we have seen, for Freud, these agencies are not defined by their location, but by their spatial relationships to one another. The super-ego becomes the chief influence in repression: this agency is capable of setting up ideals, being a moral conscience, censoring dreams and policing the borders between the conscious and the unconscious. In other words, the super-ego draws on influences from the surrounding environment, and the demands that this environment makes on the ego, to set up ego-ideals which the ego must strive to comply with, but can never achieve.

It is the (relative) strength of the anxieties induced by loss, lack and melancholy which both initiate and sustain the personality development of the child – and this occurs within a maturational environment which is profoundly social and intersubjective from the outset. Up to the age of five, the child's first love objects are its parents, but repression forces the child to renounce these objects and the child's drives become inhibited in their aim. The child has to subordinate or displace its drives into socially acceptable goals and behaviours. The ways in which the child defends itself against anxiety, forms attachments and identifies with others produce a deep archaeology of the mind which, consciously and/or unconsciously, will be seen in behaviour throughout adult life.

When in a group (of whatever kind), the adult acts on the basis of the lessons she or he learnt from these early childhood experiences: hence the descriptions of the hypnotic effect of crowds (with its evocation of childlike helplessness) or of the charisma of the leader (which casts up feelings about the powerful parent, usually as distant father-figure), feelings which not only can be traced back to hidden relationships but also hide those experiences under the cloak of the present. Groups, then, consist of a number of individuals 'who have put one and the same object in the place of their ego ideal and have consequently identified themselves with one another in their ego' (Freud, 1921: 147).

Freud has now taken this analysis of group emotional ties as far as he can; he is now tempted into discussing the idea of a 'herd instinct'. This suits his purpose because he is looking for a paradigmatic myth which can situate and explain the movement of history towards an emotionally-crippling civilisation – a myth he will also alter and complete in 'Civilization and its discontents' (1930: 288–298). Freud's foundational myth of a distinctly human psyche relies on the story of the 'primal horde': this is his model of civilisation and the group – it would be this that a distinctly Freudian critical theory would use to assess the social (see note 2 above). It is difficult not to find Freud's primal horde story bizarre – reading it, however, does help describe the limits of Freud's analysis of the group, history and civilisation.

Drawing on Darwin's theories of evolution and his own analysis of religion (1913), Freud's myth starts in the mists of time: 'the primitive form of human society was that of a horde ruled over despotically by a powerful male' (1921: 154). In order to substantiate this assumption, Freud argues that traces of this primal organisation can be found in totemism, religion and morality. Social organisation is connected to the killing of the despotic male and the conversion of the horde into a community of brothers (see also MacCannell, 1991, Chapter 1). In terms of Freud's original problem about the relationship between the individual and the group, the primal horde hypothesis gives a clear indication that there must be a group psychology and a leader psychology, thereby suggesting that even in a primitive state there is an individual psychology. The primal group is, like the church or army, bound together by libidinal ties. What distinguishes the horde from the crowd, though, is that the horde's leader has few if any ties to the horde because he need not love anybody but himself or those he needs to carry out his wishes. While Freud suggests that the psychology of the leader is narcissistic, he believes that 'civilised' group members still need the illusion that the leader loves all members of the group equally.

In the primal horde, the powerful male had the possibility of sexual satisfaction on demand and it was clear to the horde that he did not love members of the group equally: 'all of the sons knew that they were equally *persecuted* by the primal father, and *feared* him equally' (1921: 157). The lesson for Freud is that the contemporary groups are still headed by 'the dreaded primal father; the group still wishes to be governed by unrestricted force; it has an extreme passion for authority' (1921: 160). There he leaves the story – content that he has made his point about the importance of libidinal ties.

Freud completes this sorry tale in 1930, by which time the primal horde has become a site of mutual aid; where the male needs to keep a female in order to provide him with sexual satisfaction, and the female needs a strong male to look after her (1930: 288). Nevertheless the horde is still led by a despotic male with arbitrary power. Belatedly, the brothers band together and kill the primal chief because the brothers love their mothers and sisters. They set up restrictions which all the sons must now obey, because both the hostility of the sons towards the father and the rivalries of love cannot be allowed to go unchecked (1930: 302). Culture is set up over the heads of all individuals, to which all individuals must submit. In this newly-made communal life, two conflicting elements are instituted: the necessity of work (Ananke) and the power of love (Eros). The group member must work to live, but love makes the individual unwilling to work. Love must now be tamed and put in the service of group survival: indeed in the church and the army, there is no place for women as love objects (1921: 175).[7] Over time, civilisation marks the development of rules which organise relationships between individuals, families and the state: 'The final outcome should be a rule of law which all – except those who are not capable of entering a community – have contributed by a sacrifice of their instincts, and which leaves no-one – again with the same exception – at the mercy of brute force' (1930: 287).

It is clear to Freud that 'what we call our civilization is largely responsible for our misery, and that we should be much happier if we gave it up and returned to primitive conditions' (1930: 274). Despite protecting individuals against the leader, civilisation requires the renouncing or displacement of sexual drives: culture frustrates human beings, while economic relations dictate the amount of sexual liberation that can be tolerated. Civilisation legislates which sexual objects are appropriate and at what age sexual behaviours are permitted: the perverse is forbidden. There is a requirement for a single kind of sexual life for everyone, disregarding dissimilarities amongst people. Even heterosexual genital sex and procreation is further restricted by marriage rules. These boundaries, however, are commonly exceeded (for support of this particular line of argument, see Douglas, 1966):

> Furthermore, women soon come into opposition to civilization and display their retarding and restraining influence – those very women, who in the beginning, laid the foundations of civilization by the claims of their love. Women represent the interests of the family and of sexual life. The work of civilization has become increasingly the business of men, it confronts them with ever more difficult tasks and compels them to carry out instinctual sublimations of which women are little capable.
>
> (Freud, 1930: 293)

Bearing in mind that Freud was writing before the rise of fascism in Germany, his arguments and descriptions of civilisation seem remarkably astute (see Adorno, 1978); on the other hand, they also seem markedly

7 See also Theweleit, 1977 and 1978; this point is taken up in Chapter 6 below.

misogynist and racist. It is now possible to bring this section to a conclusion which will lead into the subsequent sections, each of which draw on elements of Freud's psychoanalysis which may help inform a psychoanalytically sensitive geographical imagination.

There are obvious problems with Freud's interpretation of 'the mob', problems which may say more about Freud's fear of the 'ordinary man' than crowd behaviour (see de Certeau, 1984: 2–5). Clearly, Freud's understanding is drawn from a register which is based in masculinist, bourgeois, imperialist and Judaeo-Christian terms. His myth of the primal horde contains a father and brothers, where women serve the father. Although sexual jealousy sparks the revolt against the father, women are given no place and no agency of their own. Freud legitimated these ideas using anthropological evidence drawn directly from the contact zone between colonisers and the (about to be) colonised, but he was uncharacteristically credulous of these received ideas and he was (unsurprisingly) disinterested in the processes of the colonial production of knowledge (see Chapter 6 below).

Difficult questions still hang over Freud's account: what is the ontological status of instincts and how much can they be altered, mollified or modified? What is power? What is the relationship between power and authority? Freud tends to romanticise simple states of mind – simultaneously valuing, devaluing and amalgamating children, women and primitives – yet are these 'utopic' conditions simpler, freer, more pleasure-oriented? Nevertheless, further exploration of Freud's ideas can start to provide some useful co-ordinates in the mapping of subjectivity.

The next three subsections look more closely at debates surrounding repression, drive theory and childhood because these illustrate how Freud conceptualized the ways in which the external world impinged on the individual and the ways in which the individual responded. A further section ties this chapter together by returning to the mythologies of psychoanalysis to suggest ways in which psychoanalysis can be remapped to include missing others. It is argued that Freud's ideas help to unpack and to specify the relationships between the individual, the social and the spatial. There are several reasons for this: first, because Freud does not collapse, or reduce, one site into any other; second, because he does not assume that each site is autonomous; third, because he offers specific mechanisms which define the so-called dialectic between the self and the other; fourth, because he assumes that the self–other relationship exists in specific body-ego spaces; and, fifth, because these ideas disrupt any fixed or static notions of subjectivity, spatiality or reality. Most importantly, space and place cease to be external to the individual but become the condition of the existence of subjectivity, where the subject is placed within multiple, interacting geographies of meaning, identity and power.

GROUP PRESSURES: PROHIBITIONS, REPRESSION AND THE SPLIT SELF

According to Freud (1930), civilisation is built on the renunciation of powerful drives: the purpose of life is the satisfaction of instinctual desires but reality prevents satisfaction. The external world opposes pleasure for two reasons: first, because pleasure is transient – it can only come about as a contrast between tension and the state of absolute peace; and, second, because the satisfaction of unlimited sexuality and aggression might lead to the destruction of society. Thus, the desire for pleasure threatens us with suffering. The task of avoiding suffering pushes the desire to satisfy pleasure into the background (Freud, 1930: 271). We can do three things in the face of this reality: first, repress desire; and/or, second, take refuge in fantasy; and/or, third, build a particular reality or 'civilisation'. Society is formed to curb its member's drives and is built on the denial of what we really want (Freud, 1930: 278).

For Freud, 'the essence of society is repression of the individual and the essence of the individual is repression of himself' (Brown, 1959: 3). Everybody uses defences to deny/deflect dangerous impulses; and these defences can include reversal into the impulse's opposite, repression and sublimation. Repression is one of the defences that people use to protect themselves from painful material, and they do this by keeping it away from consciousness (Freud, 1915c; 1926b). Painful material can be, for example, the representatives of unacceptable desires, unfulfilled wishes and/or feelings of lack, absence and loss. As the social world impinges on the child, it continually tries to transform its desires into acceptable expressions.

The effect of repression is to produce an internal splitting of the mind into a conscious and an unconscious: thus, 'the repressed is the prototype of the unconscious for us' (Freud, 1923: 353). The *unconscious* is not static, but has its own dynamics. It consists of wishful impulses where there is no negation of desires, no repression (Freud, 1915a: 190). Moreover, the unconscious has no conception of time or place, contradiction or reality: it is the opposite of the world of common sense and order; it is the tireless insurgent of everyday life. Because of repression, we are alienated from parts of ourselves and we can only explain ourselves with reference to an unconscious place, hidden deep inside ourselves. There is one other place in this topography of the mind – the *preconscious*, from which ideas can be admitted to the conscious with ease, but it also contains the mind's unconscious defence mechanisms. The preconscious enables communication between the unconscious and the conscious; here are conscious memory, language, and the capacity to 'read' and 'test' reality. The preconscious is the place of censorship, the gatekeeper, between the unconscious and the conscious. As a reminder: these spaces of subjectivity are to be thought of through their spatial relationships with one another, and not as fixed sites in the mind.

In this topography, because the unconscious lies outside the conscious and the preconscious, it cannot be easily controlled. Repressed ideas are dynamic:

both because they are forced into the unconscious, and because they have a motivating effect on human actions. Freud argues that the emotional energy of the repressed desire is split off and redirected elsewhere (e.g. into dreams, parapraxes, neuroses, etc.); where desire is psychic activity aimed at procuring pleasure. Repression occurs at different times in different places and in different ways; it is a continuous struggle at all levels of the mind. This map of the mind not only does not concentrate on the conscious, but is also capable of discriminating between different aspects of the unconscious. But a different kind of perspective is needed if we are to understand how the mind is set against itself.

Freud was continually working on new models of the mind. Eventually, he came to believe that the unconscious, preconscious, conscious structure was too static; later, he proposed a co-existing, but not homologous, dynamic id, ego, super-ego framework (for the ego and super-ego, see above, pp. 85–87). These two spatial frameworks do not, therefore, map neatly one onto the other (as many have suggested): instead, they should also be thought of through their spatial relationships to one another. *The id* is the impersonal source of unconscious desires: the home of the repressed and of fundamental drives. All the id is in the unconscious, but not all the unconscious is in the id; both ego and super-ego can contain unconscious material.

The ego can be formed in two ways: by the external world impinging on the pleasure-seeking id; and, by the processes of identification, where external objects are regarded as a part of the self. Indeed, the modalities of identification are the routine way in which the mind becomes differentiated; these also install the super-ego. Though the ego is intimately rooted in the 'irrational' id, it is still the place in the mind where control is exercised over external perceptions, overt behaviour, 'reason' and 'common sense'. It is this part of our selves which we think keeps us well informed about what is going on in our hearts and minds (but doesn't and can't – there are other places where things happen).

The super-ego evolves as part of the resolution of the Oedipus complex, where the child is forced to capitulate its desires under the threat of force from the father. The super-ego acts as ideal and punishment, carrot and stick: compelling obedience to an internal authority in the same way that the child has to obey an external authority. The ego strives both to appease the super-moral super-ego and to be loved by it, but it cannot escape the sense of guilt which is inflicted by the merciless law enforcer as punishment for any kind of transgression. The ego has the unenviable task of negotiating between the demands of the external world, the id and the super-ego (Freud, 1923: 397).

There has been a great deal of controversy about Freud's account of the unconscious and the structure of the mind. This debate has focused on two interrelated points: first, the constitution of the unconscious and its effects on behaviour; and, second, the structure of the psyche and its origin. The first issue relates to the extent to which biological 'instincts' can be used to explain repression and the existence of an unconscious. The second question relates to the original self, asking whether human beings are coherent unities which then become split, or whether they are inherently split, contradictory and fragmented. I would argue that it does not help to think in terms of biological instincts, but in terms of drives which are constituted psychically, spatially and socially; nor is it fruitful to posit an original, coherent, full, complete self, especially if it is seen as 'recoverable' in some way. For me, Freud's topographies of the mind are useful in so far as they help illuminate the ways in which children encounter, understand, fantasise about, explain and perform their bodies, their minds and their place in the world.

ENDURING CHILDHOOD: DESIRE, 'CASTRATION' AND SEXUALITY

As we have seen, 'civilisation' is a process of control and limitation, of coercion of the individual in the interests of the group (Brown, 1959: 9). Thus for Freud, 'human life is to be lived in the face of misery, with society as a necessary evil' (Frosh, 1987: 41). Love and hate are denounced because the energy needed to build culture cannot be wasted on sex and/or violence. Still, repression is painful: reality does not easily overcome pleasure. So, society needs a powerful medium if it is to harness the energy of sexuality and aggression for social reproduction. In our (western patriarchal capitalist) society, *the Oedipus complex* (through which patriarchal ideology is internalised) is the pivotal drama in the development of the child, and this involves the family and family relations (in which patriarchal, bourgeois and racist ideology is reproduced).[8] This point must not be lost in a geographical imagination. The Oedipus complex may not occur everywhere, but it is a useful myth because it describes both the 'force' with which infantile polymorphous pansexuality is threatened and cajoled and how the 'force' of social prohibition and sanctions are taken on board as a condition of subjectivity, in part following the spatialities of identification outlined above. The Oedipus complex can be seen as a site where the body and subjectivity are 'territorialised' according to the (il)legibility of socially-inscribed geographies of meaning, identity and power. The important point is that, while they are *meant* to take one of two (heterosexual) routes through the Oedipus complex, children will forever find their own way out.

For Freud (1905; 1913), the starting point for the Oedipus complex is the incest taboo, but the incest taboo is *not* natural. It is the incestuous desire that is natural, reflecting the free play of desire, but social taboos instigate the repression and reorganisation of desire into socially acceptable forms. Initially, the child is unaware of taboos and the girl/boy takes narcissistic pleasure in its own body. When the child becomes aware of the existence of love objects outside its own body, she/he also experiences the force of taboo and its associated power relations. As he grapples with the Oedipus crisis, the

8 This point accords with Deleuze and Guattari's understanding of the significance of the Oedipus complex (1972, 1987).

boy is forced to disown his desire for the mother, both because of the father's authority and because of the threat of castration. For Freud, the fear of castration leads the boy to repress his desires and identify with the father. This is the model for all the male's encounters with society: desire is opposed by a patriarchal authority, an authority which is internalised and a part of him. Castration 'governs the position of each person in the triangle father, mother and child; in the way it does this, it embodies the law that founds the human order itself' (Mitchell, 1982: 14). Thus, the father is the symbol of patriarchal authority and of all social authority under patriarchy. The father originates culture and sexual difference: what is allowed, what is forbidden.

Opinions differ as to the form of the Oedipus complex. The father–mother–child triad is, for Freud, a natural matrix. In contrast, for Klein, the Oedipus complex develops as a result of the child's desire to form relationships. While the child recognises people as individuals, they are still seen through the filter of fantasy. As well as Oedipal rivalry, the child becomes envious of its parents because they are seen to provide each other with the gratification that the child desires.[9] For Laplanche and Pontalis (1973) the triangle is a symbolic matrix, consisting of the child, the bearer of the law and the child's natural object. Since this is a symbolic matrix, it may not necessarily be observed in surface interactions between parent and child. From this perspective, social relations are founded on the symbolic structures of kinship and property; through the individual's encounters with those symbolic structures; and through her/his defensive identifications in which the individual incorporates the social (and spatial) organisation of experience into the self. It can be argued that the Oedipus complex, enforced by the fear of castration, directs sexuality towards acceptable expressions and models a punitive super-ego on the ideals of family, procreation (the renunciation of pleasure) and obedience to authority (see also Chapter 6 below). Freud's version of the Oedipus complex cannot, therefore, simply be taken for granted.

There is also the question of sexual difference. Until the Oedipal drama, Freud argues that there is little divergence between male and female development. Both sexes are attached to the mother, both are bisexual and so both contain the potential for masculinity and femininity. Nevertheless, for boys, during the Oedipus complex, the fear of castration forces a tremendous repression of incestuous desire, demands a strong identification with the father, and instigates a punitive and powerful super-ego. There are no such paths for girls. In Freud's understanding, the girl's Oedipus complex cannot be a simple parallel of the boy's: because she does not have a penis which can be castrated, she cannot fear castration and is therefore not subject to the consequences of this terror. Instead, unlike the boy, the girl must both renounce her own body-lacking-penis and transfer her desire from the mother to the father, partly because she wants to possess the phallus and

---

9 See, for example, Klein, 1927 and 1945.

partly because she has to repress her own hostility towards the mother.[10] In this account, it is assumed that the girl is a little man and *not* the boy a little woman: she wants 'masculinity' and hates 'femininity'. And, while castration completes the Oedipus complex for boys, it starts it for girls (i.e. when girls realise that they are already castrated, then they begin to want a penis).[11]

While the boy's erotogenesis shifts to the penis and the desire for penetration of the mother, the girl's erotogenesis shifts to the clitoris but, according to Freud (1933: 151–152), it cannot remain there – it must move again to the vagina. As the clitoris (and masturbation) is given up for the vagina, passivity is encouraged in girls. Thus, the girl, dependent on the clitoris for sexual stimulation, becomes aware of her inferiority and feels mixed and dangerous emotions – a sense of her own inferiority; her distance from power; a hateful rage at the mother for creating her in her mother's own image; and a passionate penis-envy of the father (and brother). As the girl recognises herself as already castrated, she wishes to displace the mother and share the father's power. The girl gives up wanting the mother and, instead, wants to be wanted by the father (Mitchell, 1974: 108). Finally, the desire for the penis must be renounced; it is replaced with a desire for a baby, especially a boy which brings with it the desired penis (Freud, 1933: 162). For Freud, ultimately, only men can supply women with what they really want – a (real) penis.

Undoubtedly, Freud degrades women's bodies, female sexuality, femininity and maternity. He also excludes women from culture and thinks that their anatomy is their destiny; women are 'in the grossest sense something less than boys' (Rieff, 1959: 174). Freud's invidious misogyny is insidious and pervasive; for him, women were unable to reach male potentials, marked by their predominant envy, masochism, passivity, vanity, jealousy, a limited capacity for abstract thought and a limited sense of justice.[12] Womanhood is unwholesome, perhaps because women are less repressed than men and, therefore, have weaker super-egos (Freud, 1925). Freud has not been misread: his anti-femininity is foundational and corrupting – it is also revealing. Even so, feminists have found Freud's inability to solve the riddle of femininity and his description of what they prefer to see as the consequences of patriarchy as productive and illuminating for feminist politics (for a recent contribution, see Brennan, 1992). The question has been whether Freud's accounts of sexuality are descriptive (even critical) or normative (even supportive). Either way, Freud's version of socialisation and sexual differentiation must be reconstructed (see Chodorow, 1978 and 1989). Perhaps, two places to extend this reconceptualisation to are the core mythologies of Freudian psychoanalysis: Narcissus and Oedipus.

---

10 See also Mitchell, 1974: 96.

11 This is 'penis-envy'. These are bleakly phallocentric terms of reference for sexuality and it is this supposition that Lacan pursues – see Chapter 5 below.

12 See Freud, 1939: 361; 1933: 168–169; see also de Beauvoir, 1949; Firestone, 1970; and Mitchell, 1974.

MYTH PLACED: FROM NARCISSUS AND ECHO TO OEDIPUS AND JOCASTA

Though Freud uses the primal horde myth to narrate the psychic development of civilisation and the structure of the civilised mind, Oedipus and Narcissus may be seen as the two founding allegories of psychoanalysis: they provide the models for an understanding of the key stages in the development of the individual's mind (see Freud, 1905, 1923; and 1914, 1917b respectively).[13] Freud was fascinated by these myths because they seemed to speak of the repressed desires that he was discovering in clinical situations: the position of Oedipus seemed to tell of the little boy's desire for the mother and his fear of punishment from the father; while the position of Narcissus seemed to tell of the child's falling in love with itself, a normal situation for both boys and girls, but which boys more comprehensively abandon under threat of castration when they enter the Oedipus complex.

Although it must be noted that Freud was aware throughout his life that his theory of femininity was 'incomplete and fragmentary and does not always sound friendly' (1933: 169), his reading of these myths is myopically focused on, or identified with, the male characters (see Olivier, 1980; Spivak, 1993). As a hermeneutic matrix of the child's situation, Freud takes as paradigmatic the position of the 'hero'. More than this, Freud fails to note the cultural specificity of incest taboos: i.e. their location in history and geography. In particular, he picks out Oedipus's and Jocasta's breaking of the son–mother incest taboo, but Freud considers only Oedipus's position in this (and not, for example, Jocasta's), nor does he think through other incest taboos such as the brother–sister or father–daughter or homosexual incest taboos (see Gallop, 1988: 55). For me, the point of rereading these myths is to see them as providing a set of dilemmas for the mindful characters, rather than a rigid set of roles which people mindlessly act through. Moreover, these myths describe specific constellations of relationships both between people and between people and their social and spatial situations.

Both these stories – Narcissus and Oedipus – are about sexuality; they are about the prohibitions placed on desire and the self-punishment that results from transgressing the rules. For Freud, as for Ovid, the essence of the Narcissus myth is that the 'boy-man' falls in love with his own reflection; realising that this love cannot be fulfilled, Narcissus decides to die. The moral of the story is that the child initially loves itself totally, because the child treats its own body as an object of love, as if it had no object relations outside itself (see Freud, 1911). In Freud's reading of Sophocles's *Oedipus Rex*, the key elements are that Oedipus kills Laius, while going to Thebes, where he meets and marries Laius's widow, Jocasta, who bears his children. When

13 Interestingly, there are two 'coincidences' in these myths: Tiresias predicts Narcissus's denouement, as he does later for Oedipus's; and the prediction is almost identical for each – they will be fine, if they do not come to know their self. For a detailed discussion of these myths and psychoanalysis, see Segal, 1988; in a different context, see Pile, 1994a.

Oedipus realises that he has killed his father (Laius) and married his mother (Jocasta), the sight of the truth is so offensive that he tears out his eyes and flees Thebes. The moral of the story is that the child falls in love with the Mother and feels aggressive towards the Father, but that it is *punishment* which not only makes this a traumatic situation but also necessitates that the child deals with it, in order to become an adult. The myths of Narcissus and Oedipus are etched into the spaces of the subject.

While Narcissus reflects the impossibility and megalomania of self-love, the Oedipus myth tells of the incest taboo and destruction ('castration', murder). Moreover, the myth can also be read in ways which seem to make the rules of acceptable desire god-given, to reinforce the presumption of heterosexuality and to emphasise the tragedy of 'perverse' love (through Oedipus's 'improper' love object and Narcissus's failure to recognise a 'proper' love object). For Freud, the child must, first, overcome their narcissism and, then, deal with the Oedipal situation. The most important phase in the development of an adult psyche is the Oedipus complex. The ways in which the child deals with its position as Narcissus and Oedipus will form the patterns of relationships which the child will (unknowingly) repeat throughout its life.

If these myths can in some way act as allegories for the ways in which the child is placed into dramas not of their own choosing and for the kinds of traumas the child might have to cope with, then these myths must be refigured: the consequence of this would not be to render Freud's work redundant, but it would off-centre his reading. In each of his myths, Narcissus and Oedipus, and other characters – already on stage – must be given back their place.[14] In each of the myths, Freud misses and underplays the role of women; however, simply adding women to the story does not solve the problem – the myths must be read as allegories of constellations of self–other relationships, but where the self and the other are not presumed to be stable, coherent, isolated or static things. In this retelling, the spaces of the secluded pool (in Narcissus), of hiding, journey and return (in both), and of the city (in Oedipus) are constitutive of the drama, the tragedy of subjectivity.

By remembering Echo's role in the Narcissus myth, it is possible to foreground Narcissus's failure to acknowledge others. Narcissus falls in love with himself only after rejecting and running away from the desires of Echo. Echo has her own story: her ability to speak was limited by Juno as an act of revenge for delaying her with endless chatter, while her (Juno's) husband, Jupiter, made love to nymphs. For Spivak, the position of Echo suggests that some people are only allowed to speak using the words of others, who are both more powerful and beyond their control (Spivak, 1993). Thus, if Echo marks the non-place of a femininity which is defined in terms of

14 This also applies in the primal horde myth, where the situated psychodynamics of the daughters, sisters and mothers is never considered.

phallocentric discourse, then Freud must be placed within a masculinist discourse which denies a place for the other either to be or to speak. This argument stresses Echo's lack of choice: she mechanically reproduces the last words of others, she always answers back. While accepting this reading, Echo also subverts the notion of a purely mimetic identity. Echo does not simply echo. In her encounters with Narcissus, Echo waited and determined which words to repeat: that is, Echo does have a power of speech. She is, then, the mark of a meaning to words that lie outside the intention of the speaker or the economy of meaning in which the speaker is embedded. Moreover, Narcissus and Echo's story suggests that there is a double meaning in 'reflection' which implies an internal relation between the visual and desire, between subjectivity and space, within power-laden cultural modes of signification. This story about this space of reflection will be picked up again in Chapter 5 below.

Similarly, Oedipus's relation to others turns out to be more complex than Freud's reading suggests: as Christiane Olivier puts it, 'Laius/Jocasta ... Jocasta/Oedipus ... Oedipus/Antigone and Ismene – there we have the Greek tragedy' (1980: 1). While Oedipus is the central character of his own story, there are other stories which make his tragedy possible. There is a story about his parents, and their place in the world, and their different feelings towards each other and their children. This, then, places the child differently in relation to each parent, wanting different things from each parent, and with each parent responding differently. This suggests a much more involved model of child–parent relationships. Laius's wildest fear was that he would be killed by his own flesh and blood, so he decides to kill Oedipus, but Laius's servant hides Oedipus instead. In the first instance, then, Oedipus is placed into a relation with his father which is marked by death, but they are (not) fighting over Jocasta. Jocasta, throughout, is ignored by Freud – and by Sophocles. Nevertheless, she has her own psychodynamic place in the narrative: 'Jocasta and her desire, which drives her to sleep with her own son, flesh of her flesh, with the man who has the sexual parts that she, a woman, does not' (Olivier, 1980: 1). Jocasta's desire is unaccounted for – and this tells in Freud's failure to comprehend femininity. Moreover, Jocasta and Oedipus are to have children of their own – two daughters: Antigone and Ismene. These children have their own desires and these have their own tragedies – no one in the family is to escape punishment for too much love (incest) or too much hate (murder).

Psychoanalytic mythologies suggest that there is a relationship between desire, the body and spatiality, and that particular conjunctures of these can result in punishment and compensation, delivered by a punitive, arbitrary and indisputable authority. Indeed it would appear that subjectivity is conditional on these elements: desire, the body, spatiality; authority, punishment, compensation; the unquenchable thirst for pleasure – and the limits placed on these by the situation. By reading these myths within their own terms, subjectivity is delimited almost exclusively by psychic reality for Freud, yet each myth is circumscribed by other circumstances – by the power of others – which he does not include, let alone develop, in the narrative of psychic development. Narcissus falls foul of the demand for retribution of a boy he has spurned, while Oedipus demands to know why his city suffers from ill-fortune. Freud's 'myth-placed' subjectivities are then shot through with differential power relations. So, while Freud's myths do highlight the ambivalent psychodynamics which constitute the ways in which individuals encounter others and those others encounter the individual, his analyses say little about the power relations which striate these experiences.

Without Freud's mythologies, however, it would be difficult to understand the ideas that he uses to interpret psychic reality and psychic mechanisms, or to understand the dynamics of self–other(s) encounters; so they cannot be simply abandoned. However, by reinterpreting the myths in ways which include the whole cast of characters and by re-evaluating the relative importance of both the characters in the myth and the myths themselves, it is possible to map out other co-ordinates of subjectivity. When using 'Narcissus' or 'Oedipus' as shorthand terms, they would refer to multi-layered and dynamic interactions between the self and others within multi-layered and dynamic geographies of meaning, identity and power. In this way, these myths tell of fate and of an inevitable tragedy, but instead provide a map of the psychodynamics between the personal, the social and spatial, dynamics which are marked by fantasy, pleasure and reality; by identification, incorporation and idealisation; by introjection and projection: all of which disrupts any fixed or autonomous understanding of the personal or the social or the spatial.

Freud was reading these myths as allegories of the boy's situation, but by rereading them as allegories of self–other relationships, these myths draw different maps of meaning, identity and power, which are played out through the body and through the setting of the scene. Instead of seeing these myths as universal models of sexual differentiation and personality formation, they become stories of how individuals cope with the situations with which they are presented, in conditions where they do not (fully) appreciate what is going on and where the transgression of taboos (intentionally or unintentionally) leads to punitive (self) punishment. The significance of these myths, therefore, lies in their expression both of social prohibitions and of the intensity of psychodynamics.

From this perspective, the Oedipus complex and (primary and secondary) narcissism become modes of coping with traumatic circumstances: for example, the realisation that the mother is not part of the body when she 'disappears'; the realisation that each parent and the self is prohibited as an object of love; the realisation that inappropriate emotions may be punished; the realisation of punishment; and so on. Each of these circumstances is individual, but structured differently for boys and girls, in different kinds of household conditions, in different social circumstances, and so on. Each is also circumscribed within particular settings, outside of which the story would be unimaginable: Narcissus has to be confined to an isolated pool (just as the mother–child is understood in isolation from the rest of the world); while

Oedipus has to be hidden away, and then to return unrecognised – indeed, the suspense in each drama is created by the protagonist's failure to recognise their 'true' identity, their 'true' place in the world. It can be concluded that spatial relationships – involved in the play of presence and absence, appearance and disappearance, the subject and the body, identity and place, desire and punishment – are constitutive of the reality of these myths.

CONCLUSION: TERRITORIES OF THE MIND

With reservations, especially concerning his failure to appreciate differences of race, gender and sexuality, it is worth quoting Freud:

> Each individual is a component of numerous groups, he is bound by ties of identification in many directions, and he has built up his ego upon the most various models. Each individual therefore has a share of numerous group minds – those of his race, of his class, of his creed, of his nationality, etc. – and he can also raise himself above them to the extent of having a scrap of independence and originality.
>
> (Freud, 1921: 161)

Each individual is tied by the bonds of love and hate, in many directions, to numerous groups; each forms their sense of self in relation to different models of behaviour; each has a share of many group identities. These relationships are spatial, but the subject's scrap of independence and originality comes from another space – their self. Freud's myths speak of the strength of feelings contained in the psychodynamics of these ties: so strong, indeed, that people can be impelled towards their own death or bodily mutilation (see also Chapter 6 below). While Freud has helped in understanding that the boundary between the self and the other is neither absolute, nor unchanging, nor impermeable, problems still remain concerning the structure and organisation of the social. And while Freud illuminates the psychic dimensions to power by exploring the libidinal investments within the group and between the group and the leader, the social cannot be explained in these terms alone because they do not address other dimensions of social relationships.

There is, however, much to be gained from Freud's account, for example, in terms of the spatialities of the mind, of the ambivalence and duplicity of motivations and of the cross-hatching of subjectivity by 'civilisation'. Moreover, he continually places the child/adult within matrices of meaningful relationships, where the child successively builds on its appreciation of those relationships. This is not an easy or conscious activity. The child solves and resolves awkward, traumatic situations in particular ways and the 'solution' is repeated.[15] Although the child repeats its solutions (i.e.

15 This links into Judith Butler's ideas about sexual identity being a sedimented, repeated performance (1990, 1993).

repeatedly places itself in the same place), the traumatic situation is forgotten – that is, made unconscious (i.e. trauma is displaced): hence Freud's famous phrase about 'repeating without remembering'. Individuals may feel well informed about themselves, but they have forgotten the reasons why they feel the way they do.

There is no essential self; the child becomes a subject through its responses to the circumstances it finds itself in: these are multiply determined, the result of the interactions between the internal and external worlds; they involve fantasy, pleasure and reality principles; conscious, preconscious and unconscious psychodynamics; agency in the form of super ego, ego and id; and a 'real' world of object relationships. More than this, though, Freud's ideas disrupt the cosy separation between the internal world and the external world, partly through concepts such as identification, incorporation, attachment, sublimation and so on. Without an essential self or an autonomous external world, there is not just one path along which all children travel (with girls and boys on different sides of the road). When Freud places people within mythological situations, it is in order to reveal something of the way in which the child finds a place in the world and not to find the place where children inevitably end up – after all, in therapeutic situations, Freud was continually encountering people whose failure to conform to social expectations was causing them psychic *and* physical distress. The social map is neither well read nor well travelled.

To return to Huxley's 'unmapped continents', it is possible to read Freud's spatial imagination in various ways. Freud clearly 'placed' the child in traumatic encounters (of mythic proportions), he also 'spatialised' the mind, and he also situated individuals and groups within (repeated but not remembered) libidinal ties. In his therapeutic practice, Freud often declared the aim of enabling the patient to take greater control of their drives and unconscious motivations by giving them greater awareness of the effects of these on their emotional and physical lives (for example, Freud, 1920 and 1923). This may be, then, Huxley's journey across continents – therapy as an exploration of fixed and autonomous places, an imperial colonisation of dangerous and exotic worlds. Freud, however, was only too aware of the ways in which these 'continents' move, transform themselves into other things and resist such a colonisation (see Freud, 1926b). An appropriate allegory of the mind is not a Victorian imperial geography, a meeting of the other within: empires of the mind are not to be seen in these terms. Instead, it is important to note Freud's use of many topographies, his alertness to fixity and fluidity, and his use of distinct allegories to describe specific situations. In all this, a spatial imagination (the synchronic existence of multiple, overlapping, interacting and indifferent 'things') proved peculiarly useful in turning fixed, one-dimensional and essentialist accounts of subjectivity inside out.

Freud persisted in the development of spatial imagination in his psychoanalytic discourse, but his work proved too diffuse for some. Other spatial imaginations were brought to bear by Lacan in his attempt to provide a

Freudian reading of Freud's work. Thus, Lacan was concerned to provide a systematic account of the position of Narcissus and Oedipus; these characters became exemplars of the structural positions which the child/adult is forced to adopt. In doing so, Lacan offered an account of power which provides the next step in a psychoanalytic account of the social. In the next chapter, Lacan's ideas about the mirror phase and the visual, the phallus and language are explored in order to clarify the spaces of meaning, identity and power.

5

MISPLACED

locating the 'I' in the field of the Other

The baby new to earth and sky,
    What time his tender palm is prest
    Against the circle of the breast
Has never thought that 'this is I':

But as he grows he gathers much,
    And learns the use of 'I', and 'me',
    And finds 'I am not what I see,
And other than the things I touch'.

So rounds he to a separate mind
    From whence clear memory may begin,
    As thro' the frame that binds him in
His isolation grows defined.

(A. L. Tennyson, 1850, *In Memoriam A. H. H.*, XLV)

INTRODUCTION

The last chapter ended with a discussion of Freud's use of two Greek myths: Narcissus and Oedipus. It was argued that these myths occupy a central place in the formation of Freud's developmental psychology. In this chapter, I will develop an argument which highlights Lacan's attempts to systematise these myths. Lacan always asserted that he was providing a Freudian rereading of Freud and that he was alert to the mythical content of Freud's work (Lacan, 1953: 39); however, Lacan's ironing of the creases in Freud's thought creates particular patterns – some of which usefully highlight the points of capture where power institutes subjectivity, some of which unhelpfully deny the subject a place to speak from.

The examination of the work of Lacan, in this chapter, is not intended to provide a definitive account of his arguments (even if this were possible), nor to elaborate the relationship between Freud's and Lacan's thinking (this has already been done excellently by Samuels, 1993). My readings are faithless and treasonable – partly because, for me, Lacan desiccates Freud's project by instituting an unfathomable lack at the centre of the subject, which ultimately leads Lacan to understand subjectivity through the dead spaces of mathematics-like laws and diagrams. The point of this chapter, then, is to find strengths in Lacan's arguments while leaving behind those parts of his thinking which are deterministic, essentialist and reductionist. Therefore, this chapter concentrates on Lacan's tactical use of spatial metaphors and analytic deployment of spatial relationships,[1] thereby opening up further possibilities for a psychoanalytically literate account of the production of space. In the conclusion to this part of the book, this account will be set alongside Lefebvre's analysis of the production of space which draws on psychoanalysis in highly selective, but strategically significant, ways.

The first three sections of the chapter describe some key aspects of Lacan's psychoanalysis, which (hardly coincidentally) are also constituted through a spatial understanding. Although the child's psychical development cannot be reduced to a singular Narcissistic or Oedipal outcome, it is clear that the child enters – and must deal with – its position in a Narcissistic and Oedipal drama. These myths tell of highly complex interactions and interrelationships between situated characters, but it is clear that Lacan's reading chooses to stress particular aspects. The central paradox of the Narcissus myth, in his rereading, revolves around Narcissus's relationship to his specular image, the way in which Narcissus gazes at himself and the fields of desire that form in this relationship. The organising metaphor, here, becomes 'the mirror'.

Meanwhile, rather than taking the Oedipus myth to be a set of embodied subject positions, it is seen as setting out a pattern of symbolic relationships – where the law of the father must be obeyed – which have the force to install a de-centred subjectivity, as the child enters language. The central figure in this revision is the phallus, which is not anything, but towards which all signifiers are oriented: it is the eye in the storm of meaning. It can be argued that there is a straightforward developmental account of child psychology in Lacan's work. The child moves through three stages: the real (Maternal), the imaginary (post-Maternal, but pre-Paternal) and the symbolic (Paternal). However, this would be to misconstrue Lacan's sense of the simultaneity of real, imaginary and symbolic 'spaces'.

This chapter concludes by pointing to the hidden role of fantasy, the complexities of 'the visual' and fields of desire, and the multiplicities of power. These lead into the conclusion of this part of the book, which explores psychoanalytic notions which account for the spaces of the subject and the subject of spatialities.

---

1 For example, it is not seen as an accident that Lacan argues that 'the subject is subject only from being subjected to the field of the Other, the subject proceeds from his synchronic subjection in the field of the Other' (Lacan, 1973: 188). The point here is that Lacan's account of intersubjectivity can only be understood by thinking of concepts such as 'field', 'Other' and 'synchrony' spatially. Meaning is also conceptualised spatially: thus, 'mortal meaning reveals in speech a centre exterior to language', not as 'the spatialization of the circumference' of a circle but as 'the surface aspect of a zone' of the three-dimensional torus, where 'its peripheral exteriority and its central exteriority constitute only one single region' (Lacan, 1953: 105).

REGIMES OF THE VISUAL: THE MIRROR STAGE, THE I AND THE EYE, AND THE DESIRE OF THE OTHER

Much as for Tennyson, Lacan believes that 'The baby new to earth and sky ... Has never thought that "this is I" ... And finds "I am not what I see, / And other than the things I touch". / So rounds he to a separate mind ... His isolation grows defined'. In order to account for the way in which the child will eventually be able to learn 'the use of "I" and "me"', Lacan stages an encounter between the child and its mirror image. Indeed, the mirror stage is the arena in which the child has the ability to 'recognize as such his own image in a mirror' (Lacan, 1949: 1). Sometime between the age of six and eighteen months old, the child learns to master the image of itself. Once it has realised that the image is 'empty', the child can then play with its specular image. The child experiences the relationship between its own gestures and the movements which rebound from the image, all set within the reflection's surrounding space. This experience is constituted by a complex of relationships involving a real and a virtual world, the child's body and its image, and the other people and things which are the child's setting. For Lacan, something profound is going on here: the child's jubilation with its own image 'discloses a libidinal dynamism ... as well as an ontological structure of the human world' (page 2). The mirror stage is an identification where the child transforms itself into the image as it appears to the child (or *imago*) and assumes the identity of the imago. Importantly, this dialectical form of identification is prototypical of the way in which the child will later come to function as a subject:

> This jubilant assumption of his specular image by the child ... would seem to exhibit in an exemplary situation the symbolic matrix in which the *I* is precipitated in a primordial form, before it is objectified in the dialectic of identification with the other, and before language restores to it, in the universal form, its function as subject.
>
> (Lacan, 1949: 2)

The specular image, in Freudian terms, is a kind of ego-ideal – except that Lacan does not specify the form of 'identification' other than to intimate that it involves objectification and a dialectic. In Freud's account (see Chapter 4 above), a number of possible 'dialectical' relationships could be involved: if identification, then the child is assimilating an aspect of the reflection and this transforms the child, wholly or partially, along the lines provided by the idealised image; if incorporation, then, in fantasy, the image penetrates the boundaries of the child's body and the child will keep its reflection 'inside' the body; if idealisation, the individual is elevating an aspect of the image to the place of perfection and, possibly, making this idealised object an ego-ideal; if introjection, the child, in fantasy, takes the image and its fantasised qualities from the 'outside' and places them 'inside', but in contrast to incorporation this will not involve bodily boundaries; if internalisation, the child transforms relationships with the other into 'internal' relationships thereby installing 'external' conflicts 'inside' the child's psyche. It would be easy to say that Lacan has all these possibilities in mind; however, it is clear that Lacan is working within a spatio-visual understanding of this relationship – identifications occur through the I/eye – and so Freud's embodiments of intersubjectivity have been narrowed to a particular body part. Moreover, Freud's general sense of the formation of the 'I' in relation to the child's own fantasies has been specified by Lacan as a laughable jubilation founded in a paranoid misrecognition. But I am getting ahead of Lacan's argument.

The form of the mirror stage, for Lacan,

> situates the agency of the ego, before its social determination, in a fictional direction, which will always remain irreducible for the individual alone, or rather, which will only rejoin the coming-into-being of the subject asymptomatically, whatever the success of the dialectical syntheses by which he must resolve as I his discordance with his own reality.
>
> (Lacan, 1949: 2)

The reflection provides images in an 'exterior' world which is doubled, situated in a space between the visible and the invisible.[2] The fiction of the reflection exists in a virtual world which is both there and not there – and the *imagos* (images) of this double/d world can thus present themselves in the child's everyday life as hallucination, dreams, shadows and so on. All appearances thereby take on a double meaning: showing and hiding, there and not there, situated and abstract. The fictional (and, it may be added, fantasised) patterns that present themselves to the child (in which the child has agency as it fantasises) are constitutive of the formation of the 'I': the problem for the child lies in 'the signification of space' (1949: 3). The mirror situates the child within a space, but this space has a 'de-realising' effect.[3] In the dialectic of identification with the reflection, the child learns to mimic the morphology of its fictional image and the effect of this is to both create an obsession with space and to institute an air of unreality about spatial relationships.

The child is 'captured' in space, but the 'spatial dialectic' already separates the child from the 'nature' of that space: the spatial dialectic operates through the constitutive opposition between the child's fantasy of its own spatial relationships and its specular place in the world. There is a set of geographical questions here which would disrupt Lacan's 'simple' mirror. The child is situated within a multiplicity of dialectical spatialities: the child's body, the virtual world of the mirror, the 'real' world which contains the mirror, and

---

2 Lacan's attitude towards the 'phenomenology of perception' and, later, the relationship between the visible and the invisible is conditioned by his engagement with the work of Maurice Merleau-Ponty (1945, 1964) and, to a lesser extent, Jean-Paul Sartre (1943, especially Part III).

3 In making this argument, Lacan is drawing heavily on the work of Roger Callois (1935).

the child's place in the world – and none of this is simple because any dialectic is cut across by a further dialectic of the setting: the inseparable opposition between fantasy and 'reality' effects. It ought to be stressed that this is happening prior to language and that there is no presumption here of a 'natural reality' consisting merely of brute facts. Indeed, by stressing Lacan's sense of a dialectic, it is possible to suggest a 'reality' which is constituted by – and lies beyond – both 'fantasy' and 'nature'.

The mirror stage establishes a relationship between the child and its reality, between its inner world and its external world. There is, in this relation, a primary Discord which is marked by the child's sense of its 'anatomical incompleteness' – because the child can only see from one perspective, it never has a complete sight of its body: the image simultaneously assembles and dissembles the body. The child is locked into a spatial relationship which is always (at least) 'doubled' – whole and in parts, safe and unsafe, shown and hidden:

> The *mirror stage* is a drama whose internal thrust is precipitated from insufficiency to anticipation – and which manufactures for the subject, caught up in the lure of spatial identification, the succession of phantasies that extends from a fragmented body-image to a form of its facticity that I shall call orthopaedic – and, lastly, to the assumption of the armour of an alienating identity, which will mark with its rigid structure the subject's entire mental development.
>
> (Lacan, 1949: 4)

The child is already doomed to alienation before the social stamps its mark – and this is expressed in psychic distresses such as schizophrenia (the mind-in-parts) and hysteria (the body-in-parts). The child establishes bodily 'armour' and 'rigid' ego boundaries, marking the profound schism between 'me' and 'not-me', between interior and exterior – and this provides the child with a sense of its own autonomy, but this is a profound misrecognition which is nevertheless constitutive of the child's ego.[4] In establishing its self, the child evolves a series of tactics for protecting itself from the dangers caused by the instability of its fictional place in the world:[5] reversal into opposites (for example, changing activity into passivity), inversion (turning round on the child's own self: for example, turning sadism into masochism), repression (displacing dangerous or painful material out of harm's way) and sublimation (converting unacceptable impulses into acceptable behaviour). The result is that the key structures of the child's ego are founded on a 'méconnaissance'. Usually the term 'méconnaissance' is translated as 'misrecognition' which implies that the child is duped by the mirror image. I will

4 This sense of the armouring of the body-ego will be taken up in Chapter 6 below.
5 Also see Freud, 1915b: 123 and Chapter 4 above.

substitute the phrase 'failure-to-recognise'[6] in order to suggest that the child's position is similar to Narcissus's – because Narcissus failed to recognise the desires of others, Narcissus fell in love with the reflection in the pool and, though he knew it was his face and the impossibility of his love for himself, he deliberately 'failed to recognise' his place in the world and, thus, 'chose' his fate.

The mirror stage inaugurates a dialectical relationship between the child's inner life and the exterior world, where each is constitutive of the other but where the child believes in its own individuality. Moreover, the child has also learnt to 'identify', to recognise external objects and to manipulate them and itself in order to produce specific effects: the child learns to mimic an image which it laughably takes to be its own. It is now possible for the child to enter social situations, to respond to the demands and desires of others, where the 'I' becomes an apparatus constituted by its desire for the desire of others and which is able to divert, deny and deflect its own instinctual (sexual) impulses. The child is now prepared to meet the challenges of the Oedipus complex (see next section), but this 'preparation' has already 'fixed' it into a general – though not determined, nor universal – place in the world, which for Lacan is organised through the way in which the child looks at the world.

Lacan's narrative is organised around a presumption about the primacy of a visual regime: that the child's relationship within itself, to itself and within its world is constituted through its particular, even peculiar, understanding of (failure-to-recognise) the spatial composition of what it sees. Although the mirror phase is often seen as a metaphor for the way in which the child comes to see itself as it is reflected in the reactions of others, the mirror has qualities that the other does not have – especially, the mirror is a virtual space; in any case, Lacan deals explicitly with the encounter with the other elsewhere. For Lacan, the encounter with the object (it is worth repeating that objects in psychoanalysis are objects for the whole mind, including people) is profoundly dialectical and is suspended within a visual regime: the dialectic situates the subject who vacillates between the look and the look back, where this subject position is profoundly disturbing.

A useful place to start thinking about this relationship is with an anecdote which Lacan provides of a fishing trip – known as the story of the sardine tin (Lacan, 1973: 95–96; for a related account, see Pile and Thrift, 1995b). Lacan has decided to get away from it all for a day or two, to do something which does not require 'thinking': he has decided to go fishing. On this day, he is with some fishermen from a local village. The craft is small and frail, so there is often an element of danger, and Lacan says that he is enjoying the shared jeopardy with his fellow fishermen. The moment comes to pull in the

6 This substitution is motivated by reading Freud back into Lacan; this both opens out the possibility of introducing Freud's unconscious/preconscious/conscious topology of the mind and also highlights the dynamics of fear, desire and repression in the regime of the visual.

nets, when one of the fishermen points to something floating in the sea. The object is sparkling as the sun mirrors off it. It is a sardine tin, a can which once contained the kind of fish they are trying to catch. The fisherman cries to Lacan: 'You see that can? Do you see it? Well, it doesn't see you!' Although the fisherman found the incident highly amusing, Lacan was very disturbed by this. On thinking about it, he decided that the source of this anxiety was the fact that the fisherman was wrong: the can was, in fact, looking at him, but it was the fisherman who did not see him.

Lacan is first disturbed by the blinking sardine tin, second by the fisherman's laughter. For Lacan, the common sense understanding of the situation is reversed: the sardine tin looks at Lacan, while Lacan has become invisible to the fisherman. Installed at the heart of this encounter is a two-headed primal terror: the somethingness (or agency) of the other and the dissolution of the self into nothingness. For Lacan, though, there are two kinds of 'other' in this encounter: first, the other as an object-for-the-self and, second, the Other as a moment in the exchange of meaning. The sardine tin acts in both senses: first, the can as encountered defines Lacan and, second, the can is a moment of meaningful exchange between the fisherman and Lacan, where the fisherman identifies with the can in not seeing Lacan. For Lacan, this encounter is tragic: it annihilates and terrorises him.

Lacan does not presume the innocence of the gaze and this knowledge haunts him; he insists that the lines of power flow from the gaze of an already-decentred, already-illusory subject. The regime of the visual is constituted by a dialectic between a never innocent look and its suspension within an exchange of looks – where this scene is always infused with power and meaning. The lines of power that radiate out from his eyes as he looks at the world suffer interference from the lines of power that radiate out from objects in that world. Precisely because it is infused with power and meaning, the visual cannot be understood solely in terms of the relationship between the child and its specular image. For the time being it is important to note that, for Lacan, the organising principle of both power and meaning is the phallus – where the significance of the phallus is established in and by an encounter which is visual. Thus the child, who – through the encounter with the mirror – understands visually, *sees* an anatomical difference between the sexes and this brings down the veiled phallus which covers the anxiety which this provokes (Lacan, 1958; following Freud, 1908, 1925).

In the stories about the mirror and the sardine tin, Lacan can be seen to install a space at the centre of the subject and a space between the subject and others. In both cases, the quality of this space is doubled as a gap or a distance – a lack or a split. Thus, the child becomes doubly alienated, in a particular way, from itself; where the encounter between the child and its mirror image lays the (groundless) foundations for later encounters with others. While this situation is never static because it is understood dialectically, in Lacan's schema this space is a non-space on which certain (or, arguably, all) psychic processes – such as repetition and transference – are grounded. In both stories, the encounter institutes a relationship between inside and outside, between the subject and the object: a relationship which cannot be essentialised and must be understood dialectically; a relationship which involves the subject and its place in the world.

The impossible space internal to, and constitutive of, the dialectic of the visual has other qualities. There is, for Lacan, a split between 'the image' and 'the constituting presence of the subject's gaze'. Thus, Lacan (1973, Chapter 6) suggests that there is a difference between the eye and the gaze: while the eye is the look, the gaze is the primitive institution of the look; while the eye is the look into the world, the gaze is the primitive institution of forms on the world (pages 72–73). The look into the world is regulated by the subject's expectations, movements, body, emotions and intentionality, whereas the primitive institution of forms on the world is constituted by the anxiety which the child feels at the distance between it and its image. Thus when Lacan argues that 'I see only from one point, but in my existence I am looked at from all sides' (1973: 72), he is marking the difference between the eye and the gaze.

The split between the eye and the gaze is not achieved without cost, for it is instituted by an anxiety – the threat of castration (which both correlates with Freud's account of the child's understanding of the anatomical differences between the sexes and also explains the predominance of the phallus). The gaze slides over this anxiety and escapes consciousness. In this spatial topography of the mind, the gaze always lies behind or beyond understanding – once more evoking the idea that the subject's relationship to its specular image is founded by a profound failure-to-recognise its place. Thus, Lacan looks at the sardine tin only to find himself embroiled in a primitive institution of the gaze. Lacan is looked at from all sides, yet he only looks at the can from one place. Between Lacan and the can there is an anxiety established by a constituting space which is irreducible.

Anxiety – and the threat of castration – speaks of further qualities of the visual: fantasy and desire. The child that sees its image is already fantasising about the image and its relationship to it. The child is captured by the object in front of it, thus the image becomes what the child wants to see – it is constructed out of this gaze and infused with desire. 'In this matter of the visible, everything is a trap' (1973: 93). This relationship echoes Narcissus's encounter with his image. The 'mirror' and 'tin' allegories involve the ways in which fantasy and desire are threaded through the encounter between an object (image, can), a subject (child, Lacan) and a setting (mirror, fishing). In this account, the inside and the outside are shown simultaneously to be constructed as opposites and to bleed into one another. The subject must adapt itself to the world of objects, which the subject looks at, where the objects look back. Thus, the object depends on the gaze, while the subject is also suspended in the gaze.

In the dialectic of the eye and the gaze, there is a lure and a tragedy: thus, for example, Lacan says, 'when, in love, I solicit a look, what is profoundly unsatisfying and always missing is that – *You never look at me from the place which I see you*' (1973: 103). The subject is seduced into a particular relationship between the eye and the gaze by the desire that the subject has for the desire of the other: the wish to be loved. Thus, sexuality defines the field of vision.[7] It would, however, be inaccurate to suggest that the desiring subject is fooled by the difference between the eye and the gaze, instead Lacan argues that the subject maps itself into the constitutive space between the eye and the gaze. So, while Narcissus was aware that he was looking at the image of himself and resigned himself to a particular fate,[8] the subject also finds it necessary to find a place within the field of the other (recalling Echo). The lines of sight – of both power and desire – are threads which are used to map a place in relationship to others (unlike Narcissus).

The allegories of the mirror and the sardine tin situate the subject within a relationship constituted by fantasy and desire, the eye and the gaze, the internal and the external, but these are not static categories. There is a dynamic exchange of looks which takes place within a spatio-scopic regime which defines them all. Lacan's ideas begin to undo a notion that space is somehow a passive backdrop against which bodies and subjectivity can be mapped – space looks back. Space is dynamic and active: assembling, showing, containing, blurring, hiding, defining, separating, territorialising and naming many points of capture for power, identity and meaning. While the space of vision takes a privileged place in the relationship with others – partly because the visual is integrated into the field of desire – the subject is never the centre of this scene; subjectivity is always defined against another centre: here there is a lack, a lack which privileges one signifier – the phallus.

THE CENTRALITY OF THE PHALLUS: THE THREAT OF CASTRATION, THE ENTRY INTO LANGUAGE AND DESIRE

In the mirror phase, the child has established a set of imaginary figures through which it understands (but 'fails to recognise') itself and its place in the world. This relationship is inherently spatial: it involves the child in its specular image, in virtual space, in relationship to a partial body. Following Freud, Lacan also follows the child into the Oedipus complex, although this is refigured in Lacan's work as the entry into language. This is, similarly, a moment of further alienation. The Oedipus complex, as for Freud, involves the child in a quite different pattern of spatial relationships. The child has become implicated not just in language, but also in the Symbolic space of social exchanges:

7 This position is most clearly outlined in J. Rose, 1986; the mastering masculine eye is also disclosed by Pollock, 1988a. See also G. Rose, 1993, and the introduction to this part of the book.

8 It may well be that the story of Narcissus persists precisely because it simultaneously tells of a longing for this 'lost' relationship and of the social unacceptability of this relationship and the necessity of setting it aside.

> To do this, the polar relation, by which the specular image (of the narcissistic relation) is linked as a unifier to all the imaginary elements of what is called the fragmented body, provides a couple that is prepared not only by a natural conformity of development and structure to serve as a homologue for the Mother/Child symbolic relation.
>
> (Lacan, 1959: 196)

While the child has learnt to model the world on the binary relationships present in the mirror stage, the spaces of the child now involve an 'other' register. This space, for the child, is organised through signifiers which symbolise sex. The existence of the subject must not therefore be confused with its internal space, because the subject is placed in relation to exchanges with the other which revolve around sexual imperatives – three of which are particularly significant: relation, love and procreation (Lacan, 1959: 196).

The subject becomes a reality by being placed within a set of Symbolic relationships with others and the effect of entering language, or the play of signifiers, is that the id (real), ego (ideal), super-ego topography of the psyche is established. The space of the Imaginary (installed by the mirror phase) is not, however, vacated; when the subject enters the play of signifiers, the child uses a set of imaginary figures. Another aspect of this shift of gravity from the Imaginary to the Symbolic is that Imaginary figures are organised around bi-polar relations (grounded in the Mother/Child dyad) and these figures must now be made to conform to Symbolic triads (allegorically found in the place of the Child, the Mother and Father). The child's development may thus be argued to move from monadic through dyadic to triadic patterns of relationships.

For Lacan, the Symbolic organises our experience of our desires and fantasies: and this phallus-privileging language appears to be natural and pre-given, determinate and inescapable (Cameron, 1985). But the unique male or female self, with personal desires and meanings, is an illusion. The 'Symbolic' is a set of meanings that are embedded in language: it defines a culture which lies outside the child, but within which the child has to take up a position. The truth of one's self is read from the outside in: the self is created by a series of alienations that sets the person's *position* in the maps of meaning that make her or his sense of the world, for example through incest taboos. For Lacan, the primordial Law, oriented around the prohibition of incest, is identical to the Symbolic order of language (Lacan, 1953: 65) – a view which accords with Lévi-Strauss's structuralist analysis of kinship ties.

As for Freud, Lacan sees the Oedipal drama as the site where the child is forced to organise its desires along the socially acceptable lines. In order to become a grown-up, the child is made to abandon its desires and behaviours. For Freud and Lacan, the Oedipal drama is a scene where the child takes on adult sexuality under the threat of castration – transgressive desires appear to the child to be punishable with the removal of body parts (for Freud) or the erasure of desire (for Lacan). In each of these situations, the penis (for Freud) and the phallus (for Lacan) become organising principles of adult sexuality. The implication is that Freud's understanding is different from Lacan's: thus Freud's account stresses the child's reading of anatomical difference, while Lacan sets up the phallus as a structural relationship which has no relationship to anatomy (Lacan, 1958: 282). So, while Freud stresses the child's (boy's or girl's) experience of its relationship with its mother and father, Lacan argues that the child must find a place within culture's territories of meaning and identity – and power. In the long run, however, the differences between Freud and Lacan become less stark.

For Lacan, as the child shifts between the imaginary register of the mirror and the symbolic register of language, it identifies itself with particular positions which are formed around its sense of what is 'like me'/'not like me' or 'want to be like'/'don't want to be like' and so on. But this leaves the child in a doubled world where it attempts to construct its subjectivity out of the fantasies and desires it has for the other. Thus, when Lacan (1973: 103) says that '*You never look at me from the place which I see you*' and '*what I look at is never what I wish to see*', he is not only marking the impossibility of bringing the eye and the gaze together, but also the impossibility of recovering a 'true' place through, and in, language:

> The subject goes well beyond what is experienced 'subjectively' by the individual, exactly as far as the truth he is able to attain ... this truth of his history is not all contained in his script, and yet the place is marked there by the painful shocks he feels from knowing only his own lines, and not simply there, but also in pages whose disorder gives him little comfort.
>
> (Lacan, 1953: 55)

Though speaking the lines of the script, the subject cannot know the blank page that lies beneath, while the script was written by others. Adult subjectivity is formed in relation to others ('I am (not) like that' and/or 'I (don't) want to be like that' and so on), so one's self can only be experienced as an alienated 'I' – the child cannot take the place of the other. In Lacan's structuralist analysis, language is central to this alienation: it insists on difference and provides the symbolic structures in which relationships can take place. Thus, the subject is integrated into the symbolic order by language and language produces the splitting of the mind that makes this possible (Coward and Ellis, 1977: 115). Though he is clearly drawing on the structural linguistics of de Saussure, the philosophical pragmatism of Peirce and the structural anthropology of Lévi-Strauss (until a rift between them in the mid-1950s), Lacan stresses the slippages of difference and the centrality of power. The key orientation in this map is the phallus, which is an elusive signifier – it is the gap through which the subject fails to recognise both its place in the world and the place of others. The child must however find a place in a storm of meaning, identity and power, where the eye in the storm is the phallus – to which there is no access, but around which the storm always howls.

So far, it seems, the phallus is almost godlike: it is omnipresent and the centre of power, yet moves in mysterious ways.[9] Yet Lacan's account of the phallus is an attempt to understand the ways in which symptoms and subjectivity are intertwined in the behaviour of psychotic patients. The phallus is not, then, god – although this is the kind of quality that it often takes on. The problem is that the phallus is fundamentally unknowable and multiple and as such cannot be accounted for in a straightforward way. This can be seen in Lacan's exposition of the significance of the phallus (1958) – and this will involve a discussion of desire, language and the various guises of the law.

The child may fail to recognise itself and its place in the world when it encounters the mirror, but it must cope with the threat of castration in the Oedipal situation. Without castration, there would be no reason for the child to leave the cosy lure of the mirror; without a reason to respond to the interrogations of others, it could like Narcissus remain devoted to itself or caught in a relationship between itself and its mother. The desire for the self or the mother is socially unacceptable and the child is to be cut off from it. Castrated. How the child resolves the problem of what to do with these impulses – which threaten such radical punishment – will determine how the unconscious is installed within subjectivity and which positions of subjectivity the child will adopt, especially in terms of sexuality. The child maps itself more and more precisely into the geographies of meaning, identity and power by constricting the processes and objects of identification around ideals. By deploying tactics that it has already begun to use in the mirror phase – such as reversal, inversion, repression and sublimation – the child domesticates socially unacceptable expressions of its desires.

It would be wrong to think of the mirror phase as somehow outside the social, but it is the resolution of the Oedipal situation which 'civilises' the child. Moreover, because the child is taking up positions within cultural maps of meaning, identity and power, subjectivity cannot be reduced to biology. Since the problem of castration is not resolved in relation to anatomical differences between the sexes, the question then becomes how sexuality provides different places for men and women. The problem is not easily solved by suggesting that a symbol – the phallus – organises sexuality differently for boys and girls. For, in the mirror phase, the mother is presumed to possess the phallus and it is only with the discovery of the mother's 'castration' that this idea is abandoned. Nor can the question of sexual difference be solved by arguing that the phallus operates at the level of language, while the 'thing' it both represents and disguises – namely 'sex' – is its deep organising principle such that boys and girls position themselves

---

9 Another association of this kind is between the father and God: 'the attribution of procreation to the father can only be the effect of a pure signifier, of a recognition, not of a real father, but of what religion has taught us to refer to as the Name-of-the-Father' (Lacan, 1959: 199). Similarly, Lacan boldly states that '*God is unconscious*' (1973: 59).

differently, first, in relation to sex and, subsequently and consequently, in relation to the phallus. For it is only at the stage where the genitals (penis and clitoris) become organised as site of *jouissance* that the predominance of the phallus is assured, partly because the vagina is not mapped as a site of genital penetration (Lacan, 1958: 283). Perhaps unsurprisingly, Lacan tackles the question in another psychoanalytic register – using work that had been carried out in Britain.

The answer is that there is a 'flight' from the mother by both boys and girls. Here Lacan is drawing heavily on the findings of Helene Deutsch on female masochism (1930), Karen Horney on the fear of women (1932, 1933) and Ernest Jones on the castration of desire or 'aphanisis' (1927). The problem with their work, according to Lacan, is a tendency to see the phallus as a part-object: that is, that these analysts continue to connect the symbol (phallus) and the symbolised (penis). Lacan wishes to abandon any possibility that the subject might be determined by anatomy and insists instead that the phallus is a signifier – this resolution leads Lacan into the problem of linguistics: if the phallus is a signifier, then what is a signifier? 'The signifier has an active function in determining certain effects in which the signifiable appears as submitting to its mark, by becoming through that passion the signified' (1958: 284).

The penis or clitoris may not determine anything, but the phallus has an active function as a signifier, making what it signifies submit to its name and – through desire – producing a signified which it seemingly only describes.[10] Lacan continues:

> this passion of the signifier now becomes a new dimension of the human condition in that it is not only man who speaks, but that in man and through man *it* speaks, that his nature is woven by effects in which is to be found the structure of language, of which he becomes the material, and that therefore there resounds in him ... the relation of speech.
>
> (Lacan, 1958: 284)

The 'master' of speech is actually spoken by the sign. The speaker is a set of material effects of the structure of language. For Lacan, there is no 'I' that struggles to make itself understood outside the condition of language. And this relationship is simultaneously located in space and time:

> It is the world of words that creates the world of things – the things originally confused in the *hic et nunc* of the all in the process of coming-into-being – by giving its concrete being to this essence, and its ubiquity to what has always been ... Man speaks, then, but it is because the symbol has made him man.
>
> (Lacan, 1953: 65)

10 It is this point which is so important in the work of Judith Butler (1990; 1993, especially Chapter 2) and many other post-Lacanian feminist writings.

In shifting from the child's Imaginary world of the mirror to the adult's Symbolic world of language, the child has submitted to the laws that govern language and Lacan appears to have subjected the subject to the rule of language rules: the position of the adult is determined by his or her place in language. This is not a theory which it is possible for Lacan to settle on, for his concern is with understanding symptoms (from dreams to hysteria). Symptoms (such as hysteria) are expressions of feelings which cannot be expressed through language and, so, their source cannot be language. For Lacan, symptoms are organised according to the 'laws' of another place, namely the unconscious. In this schema, the unconscious becomes a 'chain of materially unstable elements that constitutes language' (1958: 285).

Language is a system of signifiers which are related by their difference from one another, but the effects of the signifier are determined 'by the double play of combination and substitution' (1958: 285; see also 1953: 58). Here Lacan is explicitly drawing on Freud's analysis of dreams (Freud, 1900). Combination and substitution are, for Lacan, figures of speech – metaphor and metonymy – which become a topology for recognising symptoms (as they were for Freud in interpreting dreams). It is not that the unconscious is determined by language, but that 'the unconscious is structured as a language' (in one of Lacan's most famous phrases): language and the unconscious resonate with one another, but they are separate sites of subjectivity. Lacan's version of subjectivity, then, has to be understood spatially: two places determine subjectivity, language and the unconscious, but they are at odds with one another and they are distanced from each other by the phallus.

Following Freud, Lacan argues that the phallus is neither imaginary nor the organ (note: penis or clitoris) that it seemingly represents. The phallus as a signifier stands in the place of the thing that it represents. This detachment between the signifier and the signified has material effects: the meaning of an object is determined by the meaning attached to it in language. For Lacan, this gap between the signifier (phallus) and the signified (penis/clitoris) is prefigured by the gap instituted in the mirror phase. This relationship is traced through with desire. Freud's analysis of the Fort/Da ('gone'/'there') game is exemplary (Freud, 1920: 283–286). The child's continual throwing away and bringing back of the cotton reel reveals 'the ever-open gap introduced by the absence [of the mother]' (Lacan, 1953: 103–104; 1973: 62). In this example, the cotton reel becomes an object which stands in the place of the mother and is 'saturated' with the child's desires and anxieties. The signifier 'reel' is detached from the 'real' signified, the mother. It can be noted that this gap is profoundly spatial: the here/there of the reel, the here/not-here of the mother, the in-here/out-there of the image, representing both desire and anxiety.

The gap between signifier and the signified means that the speaker can never represent his or her needs in speech. Instead, the subject's needs (signifieds) are articulated as demands (signifiers) and, because there is a gap between signifieds (needs) and signifiers (demands), needs become alienated and thereby reappear as desire (Lacan, 1953: 58). In sum, the object of need, stifled by articulated demand, is presented as the object of desire. The gap between need and demand, between signifier and signified, produces desire. The phallus is the privileged signifier because it symbolises the relationship between the need for satisfaction, the demand for love and a desire founded on the impossibility of satisfaction, and not because it represents the penis or clitoris. Behind the veil which covers this gap, and towards which all things look, is the phallus and this veil also lies between the internal world (the unconscious) and the external world of others. Moreover the phallus 'can play its role only when veiled' – it must remain latent (1958: 288). The subject must bar the phallus as the signifier of the gap which creates desire, even though it constitutes and institutes the subject. And so, the subject becomes further alienated. Everything that the subject is must be repressed and the phallus gains its own language in the unconscious. The phallus is the eye in the storm of meaning, identity and power; lightning strikes are driven by the clashing of language and the unconscious.

Understood in terms of the Oedipus myth, the child is in a primordial love relationship with the mother. But the child's need to be loved by the mother is barred by the father and the child learns to transform its desires into other demands. For Lacan, these positions are structural: the threat of castration and the entry into language force the child into adopting a position in relationship to the phallus – which is the name for something which is not there but which creates desire in any case. This has the effect of organising the behaviour of the sexes, including sexuality. The phallus as privileged signifier has the effect of giving reality to the subject but simultaneously derealising, or deterritorialising, the ground on which that subjectivity is inaugurated. Thus, women relinquish femininity in the masquerade (Lacan is following the work of Joan Riviere, 1929, at this point). And, although Lacan does not explicitly say this, the same may presumably be said of men: masculinity (as with other sexual identities), then, is a masquerade without a centre or essence, but which is nevertheless presumed to have one.

To summarise the argument so far: in the Symbolic order, language has a structuring role that decentres the individual and sets the conditions for all interaction: entry into language is entry into the Oedipus complex and, because entry into the Oedipus complex involves an encounter with the power of the father to castrate, language represents the Law of the Father.[11] Gender differences created by the Oedipus complex are arbitrary and alienating. Meanwhile, sexual identities are always unstable, threatening to elide. This instability is possible because the phallus is a symbolic object onto which all meanings and differences are projected and is *not* the penis or

11 This is better understood when it is realised that Lacan is playing on the similarity between the word 'non' (No) and 'nom' (Name) in French: thus, the unconscious is founded on the father's refusal to permit the child's incestuous desires (the No of the Father), while language operates on the authority of the Name of the Father.

clitoris. Phallus is itself illusory: it is the symbolic representation of the penis. As phallus, the penis is portrayed and constituted as the originator and possessor of power in patriarchal society. During the castration complex, the phallus subjugates both boys and girls: they hope they can acquire the power of the phallus, but the phallus cannot be possessed by either. All sexuality is created by the awareness of lack of phallus.

In terms of social structures, then, the Law of the Father organises social, personal and spatial relationships in line with a particular law; in terms of symbolic structures, this is the veiled phallus, which subjugates all else. Without the signifier, no-one would know anything about anything, but through the signifier 'the subject binds himself for life to the Law' (Lacan, 1959: 199). So, for Lacan, the social world is structured according to the Law of the Father, which also organises subjectivity. Misogyny (based both on the privileging of the penis because it is misrecognised as related to the phallus and on the failure to recognise the function of the vagina in sex) creates distorted and inhibited personalities in both men and women and, in the course of doing so, reproduces itself. The Law of the Father is ratified by the threat of castration, the fear of loss of sexuality. As a defence against this aggression from/towards the bearer of the law, the child watches over anxiety in two ways: first, by identifying ideals to which it must aspire and, second, by setting up an internal watch-dog to police inappropriate behaviour (see also Freud, 1930). The policing super-ego appears as a result of the internalisations caused by the prohibitions and symbolic violence of the bearer of the law.

Lacan's radical moves are to place desire as the reason to see and to speak; to identify phallocentrism as the structure of ways of seeing and as the principle which grounds ways of communicating; and to insist that there is no essential subject, instead placing subjectivity between the unconscious and the social. Yet Lacan still presumes that the phallus is the signifier of power without explaining that power, and that the phallus (and power) is one thing. There is no clear indication when or why the phallus arrives on the scene. From the account provided above, it might be assumed that the phallus is a signifier and is therefore dependent on language. Elsewhere Lacan speaks of the phallus being a third term in the relationship between the mother and the child (1959: 197).

In so far as the mother is seen to symbolise the phallus, Lacan argues, the child identifies himself with her as the imaginary object of its desire (1959: 198). The phallus on this understanding is pre-linguistic and attached to the mother. This clearly associates the phallus with the presence of the mother rather than the absence between the child and its specular image. Questions remain as to how the phallus is detached from the mother to reappear as a signifier (other), how it comes to signify sex and lack, how the penis becomes so closely identified with it, and why the phallus alone becomes the organising principle of meaning, identity and power (so much so that it is commonly understood as a power symbol). Maybe the father (familial or symbolic) wrests the phallus from the mother, or maybe the Phallus is placed with the Father (as the creator) for safe keeping by all three parties, or maybe the unconscious and language collude to identify 'the father' with the phallus through the paternal metaphor (1959: 198 and 199). It is not clear. In any case, since there is only one eye in the storm of signification, Lacan cannot account for other sites/sights of meaning, identity and power.

Even so, the child has travelled a long way – from an internal relationship with the mother, to an internal relationship with its specular image, to full-blooded adulthood in relationship to others. Each world is spatially organised, the social constituting each scene in particular ways. The adult loses none of these scenes, but then again none is fully present. The subject maps itself into the world, but simultaneously in three registers: the real, the imaginary and the symbolic. In order to understand these spatialities, it is necessary to describe briefly these three orders.

CHANGING CENTRES: THE REAL, THE IMAGINARY AND THE SYMBOLIC, AND THE PRODUCTION OF REALITY

The child has moved from the Imaginary order to the Symbolic order, and it is in the Symbolic order that the subject is fully constituted as her or his subjectivity is organised into particular identities, along the lines especially of sexuality and gender but also of class and race and so on. Despite the centrality of the lack/phallus, here are two sites of the making of subjectivity: the Imaginary and the Symbolic. There is, however, also a third: the Real. Given that there are three orders, this might suggest that One is constituted differently in each of the three and this would tend to suggest, at least, that 'lack' or 'the phallus' or 'the subject' are in three places, or maybe even three contrary 'things'. It is necessary, at this point in the discussion, to make explicit some of these implications.

The questions which this section introduces are: what are these three orders, what are the relationships between them, and where is the subject – and what of spatiality? Since the Imaginary is set up by the mirror phase and the Symbolic is implicated in the entry into language, something of these is already known. I will deal with the Real last, partly because it is commonly held to be one of Lacan's most elusive concepts and partly because it isn't (in my reading). One more point: so far I have told a story which stresses the development of the child through stages. Such a narrative runs counter to Lacan's intentions: there is no point in the development of the child at which the real, the imaginary and the symbolic are not present. A more appropriate metaphor would be to see the child's development in terms of changing centres – like turning a ball, some part is always at the top and privileged, but its integrity is always maintained by surfaces which can only be partially seen or which cannot be seen.[12]

---

12 This image is intended to resonate with Lacan's spatial imagery (see note 1).

Alan Sheridan provides definitions for key Lacanian concepts in his translations of Lacan's work (Sheridan, 1977a, 1977b). Sheridan deals with the Imaginary, the Symbolic and the Real in the same entry. He argues that the 'imaginary' was the first concept to appear. 'The imaginary was then the world, the register, the dimension of images, conscious and unconscious, perceived or imagined' (Sheridan, 1977a: ix). In Lacan's Rome Report of 1953 (Lacan, 1953), he gives prominence to the idea of the 'symbolic'. The Symbolic is the order of signifiers, which are understood as 'differential elements, in themselves without meaning, which acquire value only in their mutual relations, and forming a closed order' (Sheridan, 1977a: ix). While the Imaginary is the order of images in relation to the ego (the 'me'), the Symbolic is the world of signifiers in relation to the subject (the 'I'). Within subjectivity, there is for Lacan a difference between 'me' and 'I'.

A third term, the Real, is linked both to the Imaginary and the Symbolic. The Real is inarticulable and therefore beyond the Symbolic, and in Lacan's later work it takes on algebraic qualities. In contrast to the fluid play of signifiers in the Symbolic and the limitless field of images in the Imaginary, the Real becomes a site of fixity within subjectivity: hence, 'the real is that which always comes back to the same place' (Lacan, 1973: 49). The Real is the site of resistance within the subject, which may be associated with images but can never be articulated.[13] Indeed, in Lacan's work, the Real became 'the ineliminable residue of all articulation, the foreclosed element, which may be approached, but never grasped' (Sheridan, 1977a: x). The Real, like Freud's 'ein anderer Schauplatz' (another scene), is an unreachable place where what is missing from the Symbolic, or for Freud what is unacceptable or traumatic, is deposited. The Real prevents the subject from being colonised by the Symbolic (by culture) or remaining within the Imaginary: 'The real is beyond the automaton, the return, the coming-back, the insistence of the signs, by which we see ourselves governed by the pleasure principle. The real is that which always lies behind the automaton' (Lacan, 1973: 53–54).

The Real, then, may be broadly equated with Freud's notion of the unconscious. For me, there are strong ties between Freud's and Lacan's

---

13 Similarly, Lefebvre argues that the unconscious is the antidote to the pseudo-subject of modern social space: it prevents the lived from being vanquished by the conceived (1974: 51; a discussion of Lefebvre's use of psychoanalysis is developed in the Conclusion to Part II). In terms which are strikingly Lacanian, Lefebvre argues that the unconscious

> designates that unique process whereby every human 'being' is formed, a process which involves reduplication, doubling, repetition at another level of the spatial body; language and imaginary/real spatiality; redundancy and surprise; learning through experience of the natural and social worlds; and the forever-compromised appropriation of a 'reality' which dominates nature by means of abstraction but which is itself dominated by the worst abstractions, the abstraction of power.
>
> (Lefebvre, 1974: 208)

For Lefebvre, the unconscious in this sense becomes the site of imaginary and real struggle with, and against, 'civilisation'.

schemata: the Real becomes the unconscious,[14] the Imaginary becomes the representation of drives and the Symbolic becomes the world of culturally-determined sanctions and prohibitions. For Freud and Lacan, a laughable 'civilisation' creates a comedy of manners and errors which are mistakenly performed as identity or personality. The human subject has no pre-existence: it is entirely created through history (and, it should be noted, spatiality), and that history (spatiality) has to be understood as a series of alienations as we are integrated into the social (and spatial) order (Poster, 1978: 89). The fictitious identification between the subject (the 'I') and the ego (the 'me') in the Narcissus/mirror phase is torn apart, as the subject positions her or his self in the Oedipal/Symbolic order, by an unconscious which is created through repression. The 'civilisation' of the subject is, then, continually threatened by the (disguised) guerrilla attacks of the unconscious/the Real. The Real is the unconscious underside to both the world of images and the world of signifiers and is, therefore, in conflict with both. However, it is only the Real and the Symbolic that are in direct opposition to one another. Thus, subjectivity becomes a site of conflict between the Real and the Symbolic – a conflict which the Symbolic *normally* wins. The subject, meanwhile, suffers multiple impulses from multiple sites which are split one from the other.

However, Lacan's account remains unsatisfactory. First, he assumes that there is a lack at the centre of subjectivity and therefore cannot account for (a) constitutive presences, (b) constitutive outsides, or (c) constitutive fantasies of lack, loss or absence. Second, Lacan assumes the existence of a universal signifier, the phallus. While Lacan insists on the instability of meaning and the fictitious core of identity, there is a sense in which meaning, power and identity are fixed by a transcendental signifier which appears from nowhere, has no history and has no place. The phallus is the sign that subjugates and marginalises women, but which no man possesses. But men have the penis, and the phallus and the penis remain inseparable – the association between the clitoris and the phallus is not pursued in Lacanian theory. Through the phallus, anatomical difference is transformed into a form of masculinist (penis) power: phallus is a material form of domination, but it is also totally illusory; so men's ability to erect their power on the reality (fantasy) of the phallus is absurd, a violence and an injustice. Lacan centres his analysis on the phallus/penis, yet this repeats this absurd injustice, and never becomes a parody.[15]

Third, similarly, Lacan assumes that, through the incest prohibition, the 'non' of the father (embodied authority) maps onto the 'Nom' of the Father (transcendental authority). By arguing that the threat of castration and the

---

14 Like Freud, Lacan separates drives from the unconscious, arguing that the Real is the accomplice of drives, much as Freud believes that the unconscious is the accomplice of the id (Lacan, 1973: 69).

15 In the way that he parodied the expressions of the hysterical woman in one of his televised lectures.

entry into language circulate around the phallus, Lacan is able to conflate castration and language, but this is an assumption – the precise relationship between the threat of castration and the entry into language is never specified. Instead, Lacan relies on the mystifying effects of the phallus and desire to weave the threat of castration onto the entry into language. At root, Lacan appears to be assuming that the (patriarchal) structures of kinship (from Lévi-Strauss's structural anthropology) and the (masculinist) structures of language (from de Saussure's structural linguistics and Peirce's pragmatism) are homologous.

Lacanian theory simply reiterates this patriarchal structure by identifying the mother with the Imaginary and the pre-social and the father with the Symbolic and the social. In this way, the mother, and by extension women and femininity, are reduced to an imaginary category: this erases their psychic lives and their role in the Oedipal and post-Oedipal organisation of the adult (see Bégoin-Guignard, 1994). So, introducing a mother that is 'social' and present throughout the child's psycho-sexual development, and having the father arrive on the scene at birth, begs questions that Lacanian theory is ill-equipped to understand because sexual difference can no longer be thought to arise simply as a unidirectional, universal shift from the Imaginary Mother to the Symbolic Father.

An account of the sexual and social development of children would also have to appreciate the role of social institutions, social practices and power relations as well as the discursive construction of subjectivity, sex, gender, the person, identity and so on (see Moore, 1994: 147). Such an argument would, further, call into question Lacan's (and Freud's) accounts of the mirror stage and the Oedipus complex. In sum, misogyny, phallocracy and patriarchy remain assumptions in Lacan's theory and so he can never sever the tie between the phallus and the penis, with the consequence that, ultimately, his account of sexual difference does rest on anatomical difference;[16] these problems have exercised feminist engagements with Lacanian psychoanalysis (for a review, see Grosz, 1990, especially Chapter 6).

Revisions of Lacan's theory are, on the surface, quite simple – it is only necessary to stress certain aspects at the expense of others. For example, if the phallus is defined only as the signifier of sex and/or power, then (a) there is no reason to presume that this is tied to the penis and (b) there is no reason to use the term 'phallus' unless it is appropriate to do so. Or 'the phallus' could be taken to be tied to the penis in many different ways, in as many ways as there are penises in fact, and this would undercut the notion of a single transcendental phallus with a 'thousand' tiny phalluses. However, these 'solutions' do not go far enough in offering an economy of love, libidinal ties and sexuality which lies outside phallocentric orientations. There is an apparently awkward double position to be marked out here: the simultaneous off-centring of the phallus to allow for not-phallocentric meaning,

16 See, for example, Cixous, 1975, Gallop, 1988, Chapter 6, and Moore, 1994.

identity and power, while still acknowledging the significance of phallic meaning, identity and power elsewhere.

In de Lauretis's account of lesbian sexuality, there is an intriguing displacement of the phallus (1994, especially Chapters 4 and 6). In part, she is interested in showing both that 'lesbian perverse desire is articulated from a fantasy of dispossession or lack of being through the personal practices that disavow it and resignify the demand for love' and simultaneously that 'the specifically sexual and representational practices of lesbianism, in providing a (new) somatic and representational ground for the work of fantasy, can effectively (re)orient the drives' (page 286). By situating this 'perverse desire' within the space of fantasy, which articulates and is articulated by an 'economy' of dispossession (castration) and disavowal, de Lauretis is able to account for a practice of love which is not phallocentric. Then again, in her theory, perverse desire does not happen in a situation where the phallus is not present – it simply isn't at the centre of things. The phallus, as I understand the implication of her argument, gains its importance as a fetish but as a fetish amongst other fetishes. So Lacan's argument (1958: 289–290) that the penis is a fetish which gains its significance in relation to the phallus (i.e. the signifier of desire) can be turned; the phallus is a fetish which gains its significance as a specific response to pre-Oedipal and Oedipal configurations of fear, desire and corporeality. It is possible that this argument could be extended to include other spaces, such as 'the mirror'.

Moreover, de Lauretis's argument resonates with other possibilities for off-centring the phallus. By suggesting that subjectivity is constituted through presences and absences (desires and losses), it can be argued that the subject maps itself into the world in complex and multiple ways – through its experience and understanding of its relationship to its body, images, the mother, the father, language, power and so on. I believe that this position links to de Lauretis's radical refusal to centre perverse desire either on the pre-Oedipal mother (against feminist yearnings for a utopic sisterhood under the sign of the mother) or on the Oedipal father (against Freudian masculinism and Lacanian phallocentrism under the sign of the father). This allows sexuality to become a cartography of the subject where there is no one path, but where specific paths are designated, sanctioned and desired. And where some stray from the path. Spatiality – as well as temporality – is not simply a passive backdrop against which subjectivity takes place, it is actively constituting the subject's relationship to themself and the world. Spatiality acts through the subject by simultaneously instituting and crossing the boundaries between same and other, between inside and outside, between the me and the I, between the eye and the gaze, and so on.

This idea is – it could be argued – also expressed in the production of space: it is possible that cities, particularly, symbolise this phallic landscape because they are impersonal, alienating, violent, and the scene of abstract visual practices (see Chapter 7 below). Cities are both spectacular and superficial (see Debord, 1967; Blazwick, 1989): they embody an illusory (real) power that subjects their inhabitants, and their spatial relations force people to conform to their rules under the threat of coercive laws. Spatiality and the production of space are not inseparable nor can they be reduced one to the other. In order to further account for the relationship between spatiality, space and subjectivity, I will stage in the conclusion to this part of the book a psychoanalytic reading of Lefebvre's account of space – partly because Lefebvre's account relies so much on his reading of psychoanalysis (especially Freud, Lacan and Kristeva), and partly because the critique which I have suggested of Freud and Lacan's writings opens up further possibilities for reconceptualising the significance of spatial relationships. In other words, replacing psychoanalysis contributes to the understanding of 'the production of space'. Before this, let me conclude by offering some conclusions concerning the value of Lacan's work for understanding subjectivity, space and power.

CONCLUDING LACAN: SOME SUCCESSES, SOME EXCESSES

For me, at least, much of Lacan's work is desiccated. He attempts to represent relationships between people and within the mind as a series of apparently fixed spatial configurations (like the internal eight or Schema R), lines and algebraic notations: a series of identifications, for example, is $i(a', a'', a''', \ldots)$; a metaphor is $f(S'/S)S \cong S(+)s$; and the field of reality in the psychotic mind is delimited by four co-ordinates – (e), (i), (I) and (M); where (e) is the ego, (i) is the specular image, (M) is the signifier of the primordial object (i.e. M is for mother) and (I) the ego-ideal. It is necessary to be cautious when using Lacan's ideas (note, for example, problems associated with using his understanding of the phallus). On the other hand, Lacan's work is an attempt to understand the unconscious meanings which underlie patients' symptoms. Like Freud, Lacan is suspicious of appearances – whether seen, spoken or taken for granted. And, like Freud, Lacan's understanding of the child's development is organised through a specially spatial imagination.

This spatial imagination offers another set of ways of understanding the location of the subject within the world. Specifically, Lacan deploys the notions of the mirror stage and of the phallus to understand the (fake) anchoring of the subject to (illusory) foundations. I have already outlined criticisms of the place I see Lacan in, but it may be useful to tell another story here. The other day, I was in a restaurant which had mirrors on the wall. On a table by one of these mirrors, a small child was (just about) standing. The child was rocking its head backwards and forwards towards the mirror, but stopping just before it hit itself against the mirror. 'Ahh. Mirror phase', I thought, 'Should I tell the mother!?' Then the child stopped. It looked into the mirror, then at the people in the restaurant, back to the mirror, back to the restaurant – and then went on with the rocking of its head. Then, it checked out the mirror and the restaurant. And so on. For Lacan, the child's encounters with the mirror, the mother, the Father, the Law and so on seem to be almost dyadic – between the child and the other. Yet, this child was looking around, placing itself, checking its world – it was already in the world. The virtual space of the mirror is special and it is different – as is the 'real' space of (in this case) the restaurant, which contained the child's mother, another woman and another (older) child, and me. The mirror situates; it is also situated: the child makes its sense of the world. The mirror trope helps, but it does not stop other things happening – in particular it does not stop other social and spatial relationships from constituting subjects.

Similarly, Lacan uses the phallus as an absent presence which organises the world of signifiers – and, in doing so, it becomes hidden. But the empty space at the centre of meaning is *presumed* to be phallic. Although this may be 'recovered' by claiming that Lacan is pathologising rather than celebrating patriarchy, critical theorists have found it difficult to escape the phallus and the effects of presuming the phallus.[17] On the other hand, Lacan's spatialisation of meaning, identity and power offers more than a simple reading of all positions as organised by a phallogocentrism. For example, if language's surface contains meaning and also hides meaning, then there are a series of consequences: what is said is not everything; what is not said is just as important as what is said; figures of speech are central to the mobilisation and territorialisation of power and identity; and power, identity and meaning can be connected up in many ways, and not just via the absent presence of the phallus.

Perhaps as importantly, language itself is not everything. Lacan was at pains to point out that there are other sites and these have their own 'language'; specifically he was talking about the unconscious. If the Symbolic is organised, spatially, through many kinds of exchanges (and not just linguistic), then the unconscious resonates with these forms of exchange as a kind of constitutive, wild underside. The notion of the phallus, instead, 'shows' that power is always present in the constitution of subjects, while the unconscious acts as a site of resistance to that constitution. An analogous spatial understanding of the Eye and the I in the fields of the Other highlights the many ways in which the child is forced to take up positions in relationship to other people and within many different social relationships.

Lacan played with other spatial metaphors when thinking about subjectivity: for example, he used the Moebius strip to talk about the direction of identifications which constitute the subject. Thus, Lacan's mapping of 'the field of reality' in Schema R suggests a relationship between 'e', 'i', 'M' and 'I' (see Lacan, 1959); he also gives these relationships direction – the vectors are ei and MI (i.e. the ego is identified with the specular image, the signifier of the primordial object is identified with the ego-ideal). This cut is mapped onto the *M*oe*bi*us strip, which is an entire surface of reality which moves between fantasy and the subject, where these are both 'inside' and 'outside'

17 See for example Irigaray, 1977, Rose, 1982 and Grosz, 1989; see de Lauretis, 1994 for a thoughtful critique of 'phallocentric' Lacanian feminism.

simultaneously. A similar kind of spatial play informs a third set of ideas in Lacan's work: the relationship between the Real, the Imaginary and the Symbolic. The content of, and relationship between, these sites has already been outlined, but what is important here is that they produce a particular understanding of spatiality.

The lived experience of space is inflected by the Real, the Imaginary and the Symbolic and – like turning a prism, or looking through a stereogram – particular aspects of this spatiality will become predominant at particular points. Spatial analysis would then be sensitive to the changing relationship between each of these constitutive orders of reality. Although there remain problems with Lacan's account of each order – for example, it may be that the Real itself changes through its resonances with the Imaginary and the Symbolic and therefore does not always return to the same place; meanwhile, the Imaginary and the Symbolic may be placed in dialectical opposition to one another, but neither having one source or one aim – thus, the terms Real, Imaginary and Symbolic may help unpack the notion of spatiality and define its changes and conflicts, attachments and identifications. And, where spatiality is seen as 'dialectical' in this sense, it may be seen to be constitutive of the subject – in its Real, Imaginary and Symbolic (dis)guises.

Finally, it is important to say something about space in Lacan's work. Space is never taken to be given, static or passive. For Lacan, there is a dialectic between the internal and the external: both of which are spatial, but space (external space) can only be placed in a dialectical relationship with the child (internal space) if both have agency. While we may disagree with Lacan's account, the mirror never simply reflects – it organises the child; at the same time, the child is making sense of – in Lacan's terms, though, misconstruing – the world around it. So Lacan teaches us that space is never less than double/d.[18] On the other hand, there is no explanation here of social relationships and, therefore, there is no account of how space is produced: for example, Lacan speaks of the predominance of geometrical space, but does not account for this. While Lacan may help understand how people come to take up particular positions within a society riddled with insidious power relationships, some work needs to be done on how those positions are themselves constituted by people. One way to go about this is to look closely at an account of the production of space which draws on psychoanalytic propositions concerning society, space and subjectivity. This will form the basis of the conclusion to this part of the book, but it will also point forward to an alternative psychoanalytically informed analysis of socio-spatial relationships.

---

18 Although Bhabha does not analyse the implications of Lacan's work for thinking spatially, his work is saturated with spatial metaphors (see Bhabha, 1994). Thus, his work might be used to exemplify the doubleness of space. In his account of the psychodrama of everyday colonial life, Bhabha thinks through the disturbing spaces in between the coloniser and the colonised which are constituted through a (never less than) double/d reflection: other than an other (Bhabha, 1990; see also Mercer, 1989). I will return to this subject in the Conclusion to this book.

CONCLUSION TO PART II
psychoanalysis and space

> Not so many years ago, the word 'space' had a strictly geometrical meaning: the idea it evoked was simply that of an empty area. In scholarly use it was generally accompanied by some epithet as 'Euclidean', 'isotropic', to 'infinite', and the general feeling was that the concept of space was ultimately a mathematical one. To speak of 'social space', therefore, would have sounded strange.
>
> (Lefebvre, 1974: 1)

In this way, Lefebvre opens his book on the production of space. Some twenty years later and in a different country, these observations do not seem out of place. While it is less unusual or strange to add the epithet 'social' to space, the concept of space which remains dominant is ultimately a mathematical one: social relations are said to be stretched out over space, but space does not stretch; or, they are compressed or distanciated in time-space, but space remains a stubbornly passive backdrop. Meanwhile to speak of 'psychic landscaping', 'the psychopathology of landscape', 'the psychodynamics of place', 'psycho-spatiality' or 'psycho-geography' would sound very strange.[1] Although these terms are not Lefebvre's, his opening arguments that little, or no, attention has been paid to what he calls 'mental space' and that space is a prerequisite for consciousness might be argued to evoke such ideas.

Chapters 4 and 5 have marked out a psychoanalytic framework through which it is possible to understand the relationship between subjectivity, spatiality and social relationships. In this Conclusion to Part II, I will develop this argument in terms of the production of space, which I will deploy in Part III in a 'psycho-spatial' analysis of the body and the city. In order to explicate this position, I will place Lefebvre's work on the production of space alongside the version I have provided of Freudian and Lacanian propositions. This three-way triangulation (Lefebvre/Freud/Lacan) will reveal something of Lefebvre's otherwise incomprehensible

1 The terms 'psychic landscaping' and 'the psychopathology of landscape' are taken from the work of Patrick Keiller and Iain Sinclair, while the phrase 'psycho-geography' occurs in the writings of the Situationist International (see I. Blazwick, in conjunction with M. Francis; P. Wollen and M. Imrie, 1989).

attitude towards psychoanalysis, but it will also enable the gaps between their respective analyses to be opened up – and in this space it is possible to articulate other possibilities for psychoanalysis and spatial understanding.

So far, I have provided a general account of children's development and of the ways in which they take up positions in the world. The 'neutrality' of this exposition is forced; it has had to be written over the top of Freud's and Lacan's stories of childhood. Their stories focus on the boy. The development of the boy is the standard against which the girl's development takes place. My description of the dramas and traumas of childhood and use of the concept of 'position' in the world paradoxically allow the child to become many genders, many sexes, many bodies, in many positions, from a starting place which cannot be assumed in advance. While such a strategy is intended to be sensitive to all kinds of difference, the problem is that it appears to forget that children are 'supposed' to be either male or female, masculine or feminine, black or white, to aspire to a 'compulsory' heterosexuality, and so on.

Psychoanalysis ought to tread the tightrope between this general explanation of difference and an exposition of the ways in which children take up their highly circumscribed places within socially-sanctioned categories of being – which are commonly grounded in, and policed through, hierarchical dualisms such as 'good' versus 'bad', 'us' versus 'them', and so on (see Sibley, 1995b; see also Pile, 1994b). Having made the point that psychoanalysis should enable an understanding of both difference and differentiation, it is now necessary to provide a psychoanalytic account of the seemingly natural, normal and inescapable categories of social distinction, but which are predicated on the power-geometrics of space.

It is now possible to take a closer look at Lefebvre's book, *The Production of Space* (1974).

PSYCHOANALYSIS AND *THE PRODUCTION OF SPACE*

Over the last twenty years or so, Henri Lefebvre's *The Production of Space* has proved to be inspirational for a number of geographers working within a broadly (and also narrowly orthodox) Marxist framework (Lefebvre, 1974).[2] Rather less note has been taken of Lefebvre's use and abuse of psychoanalysis in his arguments about spatial architectonics or his description of the visual and phallic space: a notable exception is Soja's consideration of spatialities (1989). Much more recently, two geographers have gone back to Lefebvre's work in order to examine the often hidden references to, and arguments with, psychoanalysis, see Gregory, 1995, and Blum and Nast, forthcoming. In particular, their work concentrates on reassessing Lefebvre's work in the light of an understanding of basic Lacanian tenets (see Chapter 5 above). Thus, Lacan's account of the mirror phase is compared to

---

2 For a review of this reading of Lefebvre, see Gregory, 1994, especially Chapter 6.

Lefebvre's use of the mirror as one exemplar of the visual and, similarly, their different conceptualisations of the relationship between 'the real' (generally understood as the material) and 'the metaphorical' (generally understood as the linguistic) is explicated. These geographers' conclusions, however, are intriguingly opposed.

Gregory argues that Lefebvre's commitment to changing the world, while problematic in certain respects, offers a place for radical politics that is denied by, and therefore unacceptable in, Lacanian psychoanalysis. Meanwhile, Blum and Nast unerringly point to the starkly misogynist and heterosexist assumptions shared by both Lefebvre and Lacan; they argue that these run so deep that neither can provide an adequate understanding of the multiplicities of (feminine and feminist) space. I have no quarrel with these conclusions but I also think that it is possible to read out of the gap between Lacan and Lefebvre, and out of Lefebvre's more sympathetic use of Freudian psychoanalysis, the possibility of opening up a 'space' beyond the limits of their (masculinist) understandings. Lefebvre was, after all, implacably opposed to masculine space (see, for example, pages 380 and 410).

In this section, I will assume something rather bizarre – that the differences between Freud, Lacan and Lefebvre will be productive. This is bizarre because most will assume that there are no similarities which could provide co-ordinates for mapping differences. However, the work of Gregory and Blum and Nast has partly been motivated by the need to bring out into the open the similarities that have been for too long overlooked. There is, I believe, a productive gap between Freud's, Lacan's and Lefebvre's understandings of space in which it is possible to site an argument which stresses the difficulties of subjectivity, spatiality and power, although Lefebvre's haste to construct an argument of grand historical sweep leads to a somewhat arbitrary 'take it or leave it' attitude towards psychoanalytic propositions which prevents him from providing an account of the 'subjection' of space: usually, he 'takes' Freud and 'leaves' Lacan.

The reading of Lefebvre which I provide is organised around the Lacanian co-ordinates of the visual, the phallus and the relationship between 'reality', the imagined and the metaphorical. First, however, something must be said of Lefebvre's deployment of psychoanalysis (and not just Lacan) in order to make this juxtaposition seem less contrived. From this discussion, it is possible to open out other ways in which psychoanalytic concepts can inform the production of space. Together, these ideas will be explored in the final part of this book on the body and the city, the introduction to which will develop this argument further.

Introduction: Lefebvre's ambivalence towards psychoanalysis

Lefebvre's *The Production of Space* is a complex book, containing many themes, many layers, many analytic co-ordinates and a grand historical vision. The point of the book is to make an argument about the necessity of including space in any theory of a socialist transformation of society: 'To change life ... we must first change space' (page 190; see also page 419). Accordingly, Lefebvre draws on, and picks arguments with, many prominent French theorists of the time. Most commonly in geography, the historical materialist elements of Lefebvre's work have been extracted to underpin an argument about the importance of geography not only for resolving contradictions within capitalism but also for radical (actually, historical materialist) social theory. I will leave these arguments to one side, partly because I want to take them as read and partly because I am also making a one-sided and partial extraction – of the bits where Lefebvre talks about psychoanalysis, psychoanalysts, Ego, the unconscious, sexuality, desire, mirrors, the phallus and language – to point to a different explication of the 'dialectic' between social production of space and the spatial production of the social. This reading suggests that Lefebvre's basic understanding of space is profoundly psychoanalytic: thus, for example, even in his attempts to transcend the metaphor, 'the mirror' remains paradigmatic in Lefebvre's spatial architectonics. Nor is this just a theoretical sideshow for, as Lefebvre intriguingly suggests, 'to change life ... we must first change space. Absolute revolution is our self-image and our mirage – as seen through the mirror of absolute (political) space' (page 190).

Perhaps a good place to start this discussion is with his history of space – a gloomy appraisal, which Lefebvre uses to strengthen his argument for radical change – because this will open up the question of different spaces.

Lefebvre argues that there have been distinct phases in the history of space. Where once there was 'absolute space', now there is 'abstract space'. 'Absolute space' existed as fragments of nature in a time when people were intimately bound up with that space; however this natural space was invaded and appropriated. Now, people have been abstracted from this natural relationship (see pages 48–49).

> Thus social space emerged from the earth and evolved, thanks to a stubbornly pursued process of 'intellectualization', until an abstract space was constructed, a geometric, visual and phallic space that went beyond spatiality by becoming the production of a homogeneous and pathogenic political 'medium' at once aberrational and norm-bound, coercive and rationalized: the 'medium' of the state, of power and its strategies.
>
> (Lefebvre, 1974: 377)

There are many elements in this passage: that abstract space has geometric, visual and phallic 'formants' ought to be enough to mark out psychoanalytic (i.e. Lacanian) concerns; and the idea that space is a medium of power bound up not only with the state and coercion but also with norms and rationality resonates strongly with the psychoanalytically-informed analyses of the Frankfurt School; but there is also here 'a stubbornly pursued process of "intellectualization"' – the implication in the extract is that the emergence of abstract space was caused ('constructed') by a process of intellectualisation.

Similarly, Lefebvre argues that there has been a separation between knowledge and social practice which has left only relics, 'words, images, metaphors' (page 25). The question which occurs to me is this: if intellectualisation is a *mental* process, and therefore open to psychoanalytic understandings, then in what sense could it be said to *cause* social space 'beyond spatiality'? In looking at this issue, I will show that psychoanalysis is never far from the surface of these conceptual grids.

In some ways, it is not difficult to prove that Lefebvre uses psychoanalysis: there are references throughout to 'psychoanalytic topographies' (page 3) and the 'psychoanalysis of space' (pages 99, 242 and 315), while psychoanalysis has 19 entries in the index, the Ego has twelve, Kristeva and Lacan four each and Freud one. However, this kind of content analysis does not convey anything of the significance of psychoanalysis in Lefebvre's conceptual understanding of space – after all, terms like mirror, phallus and desire are not exclusively psychoanalytic. Before turning to 'the mirror' and to 'the phallus', there are two aspects of Lefebvre's spatial architectonics which I would like to suggest resonate with psychoanalysis: first, Lefebvre's understanding of social prohibition; and, second, his three-fold division of space. Let's return to the beginning of Lefebvre's work, where he sets out the conceptual grids through which he interrogates the production of space, which will lead into a discussion of the importance of space in the constitution of social sanctions. The next subsection, 'Society and space', will deal with Lefebvre's triadic understanding of spatial architectonics.

In the first few pages of *The Production of Space*, Lefebvre sets out an argument that suggests that there is a gap between different kinds of space: these are linguistic mental space, the space of social practice and the real space of conceived essences (page 5). Lefebvre argues that there is a need to conceptualise the specific links between these spaces and so he criticises Kristeva, Derrida and Barthes for their sophistry 'whereby the philosophico-epistemological notion of space is fetishized and the mental realm comes to envelop the social and physical ones' (page 5). The problem, for Lefebvre, is that there is a conflation of the mental and the social.[3] Without further comment, Lefebvre backs up this claim in note 11 by referring to Barthes talking about Lacan in this way: 'his topology does not concern *within* and *without*, even less *above* and *below*; it concerns, rather, a reverse and an obverse in constant motion – a front and back forever changing places as they revolve around something which is in the process of transformation, and which indeed, to begin with, *is not*' (Barthes, 1966: 27; cited by Lefebvre, 1974: 5; see also below). Lefebvre's objection would appear to be that Lacan's constant reversal of front and back is a constant and unmediated reversal of the mental and the social. This runs counter to Lefebvre's argument that

3 Although Lefebvre admits in note 12 (page 5) that both Derrida and Kristeva acknowledge that there is an intermediate articulation between the mental and the social – in Derrida's case by 'writing' and in Kristeva's case by the body.

there are different spaces, with specific links between them.[4]

Lefebvre argues that every mode of production has its own space: thus, for example, in the case of Rome or Athens, 'the ancient city had its own spatial practice' (page 31). By way of elaboration, this means that the social space of the mode of production involves both the social relations of production and the relations of reproduction. Thus, social space is involved in the reproduction of the mode of production in three key sites – the family, labour power and the social relations of production – because space contains not only the double or triple interactions between these sites, and also because it contains representations of these social relations. Moreover, social space tends to maintain the mode of production by representing social relations in a way which maintains their coherence and their cohesiveness (page 32).

Following this line of argument and, it might be added, in a moment of psychoanalytic and anthropological exuberance, Lefebvre states that:

> Representations of the relations of reproduction are sexual symbols, symbols of male and female, sometimes accompanied, sometimes not, by symbols of age . . . relations of reproduction are divided into frontal, public, overt – and hence coded – relations on the one hand, and, on the other, covert, clandestine and repressed relations which, precisely because they are repressed, characterize transgressions related not so much to sex *per se* as to sexual pleasure, its preconditions and consequences.
>
> (Lefebvre, 1974: 32–33)

If social space is about not only the relations of production but also reproduction, then from this passage it is clear that psychoanalytic processes underwrite socio-spatial reproduction. It is also apparent that a Freudian and Lacanian register is being used to interpret social relations of reproduction: it is about repression (Freud), about sex and sexual pleasure in an economy of social prohibition and sanction (Freud) and about the Symbolic order (Lacan). It could be noted that social space relies on the Symbolic order; better, it could be said that social space constitutes, and is constituted by, the Symbolic order; and it might be argued that notions of transgression, repression, overt and covert, clandestine and public, recognise social power relations – which are mapped out in city spaces, whether ancient or modern (aligning pages 31 and 33 – I will return to this point in Chapter 7 below). In Lefebvre's schema so far 'reproduction' is about the family, gender differentiation and the Symbolic order, and is therefore open to psychoanalytic explanation. However, Lefebvre wishes to dissociate himself from crude

---

4 It should be said that Lefebvre's disparaging use of Barthes's interpretation of what appears to be Lacan's understanding of the phallus (to which Lefebvre will return) is overdrawn: partly because of the way Lefebvre's ideas relate to Barthes's version of Lacan's phallus and partly because of the apparent correspondence between Lefebvre's three spaces and Lacan's triadic topology of the Real, the Imaginary and the Symbolic (see below).

psychoanalytic and anthropological interpretations of the social.[5] Thus, he complains

> some would doubtless argue that the ultimate foundation of social space is *prohibition*, adducing in support of this thesis the unsaid in communication between the members of a society; the gulf between them, their bodies and consciousness, and the difficulties of social intercourse; the dislocation of their most intimate relationships (such as the child's with its mother), and even the dislocation of bodily integrity; and, lastly, the never fully achieved restoration of these relations in an 'environment' made up of a series of zones defined by interdictions and bans.
>
> (1974: 35)

Lefebvre elaborates this statement in psychoanalytic terms: the prohibition involved here is not only that which prevents incest between the child and its mother, but also the prohibition which dislocates and fragments the body as it is represented in language (i.e. where language tends to fragment the body into the names of body-parts).[6] In a (Freudian and Lacanian) psychoanalytic moment, he argues that through the incest taboo

> the (male) child suffers symbolic castration and his own phallus is objectified for him as part of outside reality. Hence the Mother, her sex and her blood, are relegated to the realm of the cursed and the sacred – along with sexual pleasure, which is thus rendered both fascinating and inaccessible.
>
> (1974: 35–36)

Nevertheless he warns (against Lacan, but not necessarily against Freud) that, by locating castration in the symbolic and by relegating the mother's body to a polarised economy of meaning, this approach prioritises 'language over space' (page 36; see also page 136). For Lefebvre, Lacan's argument that words speak people amounts to the linguistic determination of socio-spatial relationships, while Freud's assumption that the incest taboo is the origin of civilisation fails to take into account productive activity. Again, Lefebvre is arguing that psychoanalysis only provides an account of reproduction and not of production and is therefore insufficient as an explanation of space. He pushes this point further:

> The existence of an objective, neutral and empty space is simply taken as read, and only the space of speech (and writing) is dealt with as

---

5 Lefebvre is suggesting here that Lévi-Strauss's use of psychoanalysis leads him to see kinship relationships as entirely organised around incest taboos (after Lévi-Strauss, 1949).

6 This analysis will strike some as resembling Deleuze and Guattari's identification of 'bodies without organs' (1972, 1987); indeed, they draw on Lefebvre in this context (1972: 251), while Lefebvre cites *Anti-Oedipus* (1974: 22), but he does not accept Deleuze and Guattari's account of desire (see 1974: 200).

> something that must be created. These assumptions obviously cannot become the basis for an adequate account of social/spatial practice. They apply only to an imaginary society, an ideal type . . . All the same, the existence within space of *phallic verticality*, which has a long history but which at present is becoming more prevalent, cries out for explanation.
>
> (1974: 36)

Lefebvre resists psychoanalytic versions of space because, he argues, they assume space to be passive, fixed and undialectical and because, he believes, psychoanalysis is only useful for interpreting the social relations of reproduction and representational spaces,[7] but he also paradoxically cries out for an explanation of space, of phallic verticality, which begs for psychoanalysis. This ambivalence is revealed later in his work,[8] where he is much more receptive to the idea that prohibition founds social space:

> That space signifies is incontestable. But what it signifies is dos and don'ts – and this brings us back to power. Power's message is invariably confused – deliberately so; dissimulation is necessarily part of any message from power. Thus, space indeed 'speaks' – but it does not tell all. Above all, it prohibits.
>
> (1974: 142)

There is an argument here about the relationship between power, language and space which runs counter to Lacan's by suggesting that space speaks and

---

7 Lefebvre's belief in psychoanalysis's usefulness for thinking about representational spaces is clear when he argues that:

> Representational space is alive: it speaks. It has an affective kernel or centre: Ego, bed, bedroom, dwelling, house; or, square, church, graveyard. It embraces the loci of passion, of action and of lived situations, and thus immediately implies time. Consequently it may be qualified in various ways: it may be directional, situational or relational, because it is essentially qualitative, fluid and dynamic.
>
> (1974: 42)

Moreover, Lefebvre argues that representational space, *lived* space, has its origin in childhood, with its hardships, achievements and lacks (page 362).

8 For example, Lefebvre continues:

> It is true that explaining everything in psychoanalytic terms, in terms of the unconscious, can only lead to an intolerable reductionism and dogmatism; the same goes for the overestimation of the 'structural'. Yet structures do exist, and there is such a thing as the 'unconscious' . . . If it turned out, for instance, that every society, and particularly (for our purposes) the city, had an underground and repressed life, and hence an 'unconscious' of its own, there can be no doubt that interest in psychoanalysis, at present on the decline, would get a new lease of life.
>
> (1974: 36)

Of course, it may be argued that every city – in all sorts of ways – has an underground and repressed life and that this is open to psychoanalytic explanation (see for example Williams, 1990). On the other hand, it may be that this version of 'unconscious space' is too literal (see the discussion of Real spatiality below). Lefebvre's use of psychoanalysis in the interpretation of urban life will be explored in greater detail in Chapter 7 below.

that power speaks; on the other hand, Lefebvre appears to be saying that space, above all, prohibits – it is both the medium and the message of power. Indeed, in what might be claimed as a Lacanian mood, Lefebvre argues that the Law and space are bound up one in the other, and that space prohibits and sanctions bodily behaviour. The dialectic is this: prohibition – as power – produces space and space – as prohibition – produces power. Thus, 'there can be no question but that social space is the locus of prohibition, for it is shot through with both prohibitions and their counterparts, prescriptions' (page 201).

In sum, Lefebvre's attitude to psychoanalysis is thoroughly ambivalent. He is scathing about psychoanalysis when it appears to explain the social in terms of the psyche; when it appears to explain the spatial in terms of the linguistic;[9] and when it offers one-dimensional interpretations of the social.[10] On the other hand, Lefebvre is prepared to appropriate psychoanalytic terms of reference in his analysis of society and space. Social space, then, prohibits and prescribes in ways which can be understood psychoanalytically – involving a narrative about the phallus, castration and the Mother, but Lefebvre also argues that there is more to this story.

Society and space

Lefebvre suggests that there are three social spaces – linguistic mental space, the space of social practice and the real space of conceived essences (page 5): these are further specified in order not only to specify the content of each and the relationships between them but also to reveal the power relationships inherent in each location. Lefebvre's classification of social space can be compared to, and contrasted with, Freudian and Lacanian matrices. Lefebvre's schema is well worth repeating at length (the definitions are assembled from page 33 and pages 38–39; versions (a) and (b) respectively):

1a *Spatial practice*, which embraces production and reproduction, and the particular locations and spatial sets characteristic of each formation. Spatial practice ensures continuity and some degree of cohesion. In

---

9 For example, Lefebvre resists a 'pessimistic' attitude to signs which holds that space is deadly: deadly because it is the locus of prohibitions, castration, the death instinct and the objectification of the phallus – Lefebvre rescues space by separating it from language (pages 135–136).

10 For example, Lefebvre derides Freud's 'bio-energetic' theories of the mind because they tend to become mechanistic. For Lefebvre, Freud's distinctions between Eros and Thanatos, and pleasure and reality, become interplays between fixed entities, loosing their dialectical character, and becoming little more than metaphors for the reduction of energy imbalance in the mind (Lefebvre, 1974: 177–178 and 180; see for example Freud, 1917b and 1920). Not only does Lefebvre forget that this is only one of Freud's maps of the psyche, elsewhere he deploys (an explicitly dialectical understanding of) Freud's bio-energetics and the destructive force of the death instinct (see page 180). Similarly, Lefebvre suggests that Freud's Oedipal triangle is also mechanistic and cannot account for diverse outcomes (pages 248–249). Once more, it is questionable whether this caricature of Freud's position is useful; partly, at least, because it is possible to see these ideas as part of an interpretative and dialectical, rather than mechanistic, framework (see Ricoeur, 1970; see also Chapter 4 above).

terms of social space, and of each member of a given society's relationship to that space, this cohesion implies a guaranteed level of competence and a specific level of performance.[11]

1b *Spatial practice*: the spatial practice of a society secretes that society's space; it propounds and presupposes it, in a dialectical interaction; it produces it slowly and surely as it masters and appropriates it.

2a *Representations of space*, which are tied to the relations of production and to the 'order' which those relations impose, and hence to knowledge, to signs, to codes, and to 'frontal' relations.

2b *Representations of space*: this is the dominant space in any society (or mode of production). Conceptions of space tend, with certain exceptions ... towards a system of verbal (and therefore intellectually worked out) signs.

3a *Representational spaces*, embodying complex symbolisms, sometimes coded, sometimes not, linked to the clandestine or underground side of social life, as also art (which may come eventually to be defined less as a code of space than as a code of representational space).

3b *Representational spaces*: space as directly *lived* through its associated images and symbols, and hence the space of 'inhabitants and users' ... This is the dominated – and hence passively experienced – space which the imagination seeks to change and appropriate. It overlays physical space, making symbolic use of its objects. Thus representational spaces may be said, though again with certain exceptions, to tend towards a more or less coherent system of non-verbal symbols and signs.

I admit that when I first saw this (version (a) rather than (b)), I immediately thought of Lacan's Real, Imaginary and Symbolic orders. The similarities seemed clear: '*spatial practice*' was the structure of socio-spatial relationships which underlie social space (Lefebvre) and was thus the other place ('ein anderer Schauplatz') which was the 'unconscious' or 'Real' of social relationships (after Freud and Lacan); '*representations of space*' (Lefebvre) was about images and therefore about 'the Imaginary' (after Lacan); while '*representational space*' (Lefebvre) was about symbolism and therefore about 'the Symbolic order' (after Lacan). The feeling that the similarity between Lefebvre's and Lacan's conceptual grids was more than a coincidence grew stronger while reading Pollock's analysis of modernity, women's spaces and the spaces of femininity (1988b). In making a case for feminist art history, Pollock argues that spatial arrangements and social processes are linked. She points out that the relationship between space and society concerns 'not only the space represented, or the spaces *of* the representation, but the social spaces from which the representation is made and its reciprocal positionalities (1988b: 66). Pollock's position is feminist;

11 Lefebvre states in note 31 (page 33) that he has 'borrowed' these notions of competence and performance from Noam Chomsky. It should be possible to develop both these notions in other – psychoanalytic – ways by following the work of Judith Butler, in particular, 1990.

although distinctly her own, the analysis draws implicitly and explicitly on Lacan; yet the terms loudly echo Lefebvre's – but Pollock neither alludes to nor cites Lefebvre. I speculated that the source of the isomorphism between Pollock's and Lefebvre's matrices lay in their critical engagement with (mainly Lacanian, but also Freudian) psychoanalysis. Further, when I placed Freud, Lacan and Lefebvre side by side, a whole series of triads and transition points started to appear: a table could be constructed (see Table 1) – although I should warn that the point of this exercise, now, is to move beyond it rather than to (falsely) correlate one schema with others.

Table 1 Triads and transitions in Freud, Lacan and Lefebvre

Freud		Narcissus		Oedipus	
Freud[12]	Unconscious		Preconscious		Conscious
Lacan[13]		Mirror		Phallic	
Lacan	Real		Imaginary		Symbolic
Lefebvre	Natural		Absolute space		Abstract space
Lefebvre	Perceived		Conceived		Lived
Lefebvre	Spatial practice		Representation of space		Represent-ational space

This is too easy. For example, in Lefebvre's definition of 'representations of space' on page 33, it appears that he is talking of images and therefore of the Imaginary, but by pages 38–39 it seems that he is talking of verbal exchanges and therefore of the Symbolic: thus, in Lefebvre's latter exposition, both 'representations of space' and 'representational spaces' may be claimed to be aspects of the Symbolic, while any Imaginary order disappears. Moreover, it would be nice to presume 'vertical' equivalences, thus: unconscious is Real is perceived is spatial practice. Unfortunately, I do not believe that this can be done without damaging each of these ideas – the transcoding of psychoanalysis and space has to be wary of the violence of abstraction. It is not so much that these similarities don't exist, but that there are distinct and important differences between these schemata. Though both Lefebvre and Lacan have a sense of dialectics, they play this out differently in their schemes. Thus, Lefebvre generally seems to oppose spatial practice (thesis) and representations of space (antithesis) in the production of lived, representational, space (synthesis) – although he also argues that each may

12 There are of course other Freudian schemes, such as pleasure/reality and oral/anal/genital, but the implications of these for Lefebvrian thought will have to wait.

13 It may be objected that there is no place for 'Lack' in these representations, but I believe that 'lack', the 'object petit a' and the optical unconscious should be seen as constituents of the Real, and Imaginary and the mirror phase.

contribute in different ways, depending on the mode of production (page 46); while Lacan sees the Real, the Imaginary and the Symbolic as being in dialectical opposition to each other – each having its own dialectic, which not only propels the child through the stages but also does not allow the child to 'out live' or synthesise them.

On the other hand, some kind of transcoding does seem appropriate: i.e. I believe that the similarities between these schemata are about more than a meaningless or superficial resemblance. So it seems possible to rewrite their understanding of dialectical relations in ways which are compatible. Similarly, Blum and Nast (forthcoming) reveal the similarities between the schemata of Lefebvre and Lacan; in their article, they correlate Lefebvre's Natural/Absolute/Abstract spaces with Lacan's Real/Imaginary/Symbolic orders. Moreover, there is the question of language in both Lacan's and Lefebvre's work: as Gregory (1995) points out, both Lefebvre and Lacan have a similar appreciation and suspicion of language; both play on the slash between signifier/signified; both draw on the distinction between metaphor and metonymy (for a full discussion, see Gregory, 1995; see also below), which may be likened to Freud's distinction between condensation and displacement, respectively.[14] Given this, it is possible to rewrite Lefebvre's versions of spatial practice, representations of space and representational spaces along Lacanian lines thus:

1 *(Real) Spatial practice* embraces not only the organising principles of production, reproduction and consumption which form the unconscious (in its topographical and dynamic senses) of society, but also the particular locations and spatial sets characteristic of each society. Because spatial practice always returns to the same place, it not only ensures continuity and some degree of cohesion, but is also constituted through contradictory tendencies for fragmentation and disintegration.[15]

---

14 Freud (1900) uses the distinction between 'condensation' and 'displacement' in the analysis of dream-work. Condensation is similar to metaphor in that manifest meaning is 'smaller' than latent meaning; while displacement is similar to metonymy in that one idea is substituted for another idea that is associated with it. Freud, it should be added, also used two other notions: 'representation' (the transposing of the dream into language) and 'secondary revision' (the revision of the dream in the recounting of it).

15 Elsewhere Lefebvre states that:

> Spatial practice regulates life – it does not create it. Space has no power 'in itself', nor does space as such determine spatial contradictions. These are contradictions of society – contradictions between one thing and another within society, as for example between the forces and relations of production – that emerge in space, at the level of space, and so engender the contradictions of space.
>
> (1974: 358)

While this sense of spatial practice installs the notion of contradiction at its heart, it also undervalues the spatial constitution of social practice. More promisingly, Lefebvre also argues that spatial practice is not just urban, economic or political, but also theatrical, erotic, ambiguous, full of needs and desires; and thus, in a psychoanalytically guided understanding, that:

2 (*Imaginary*) *Representation of spaces* is tied to spatial practices, but exists as the ways that those spatial practices are represented; this is the realm of images – conscious and/or unconscious, perceived and/or imagined – and hence of knowledge, of signs and codes.
3 (*Symbolic*) *Representational spaces* embody complex symbolisms, which have conscious and unconscious resonances and are therefore linked to 'underground social life', which take on meaning only in the process of exchange and as part of a system of differences; and, because meaning is associated with value, it is here that power relations are at their most visible.

While the idea of Imaginary spatiality looks pretty much like Lefebvre's representation of spaces, Real spatiality and Symbolic spatiality are quite different (sadly, damage has been done). The relationship between them must be altered as well. Real spatiality is the unconscious (as set of contradictory organising principles)[16] underside to both Imaginary spatiality (the world of images) and Symbolic spatiality (the world of signifiers) and is, in this framework, in conflict with both. However, it is only Real spatiality and Symbolic spatiality that are in opposition to one another; Imaginary spatiality both lies between and to one side of the other two spatialities; this accords with Lefebvre's analysis of 'the world of images and signs' (page 389). Thus, subjectivity becomes a site of conflict between Real spatiality and Symbolic spatiality – a conflict which Symbolic spatiality *normally* wins; i.e. it is the Symbolic that tends to deny/deflect the contradictory tendencies within Real spatiality, and Real spatiality which carries out a guerrilla warfare with Symbolic spatiality.[17] The issues here are that Real spatiality is inaccessible, except through Imaginary and Symbolic spatialities (again, see page 389); that there are dialectics within each spatiality and between them; that no one spatiality can be said to be determinate; and, finally, that no spatiality can be said to be more real or more material than any other. All this takes 'the production of space' out of a truly Lefebvrian register; it

---

> Space is liable to be eroticized and restored to ambiguity, to the common birthplace of needs and desires, by means of music, by means of differential systems and valorizations which overwhelm the strict localization of needs and desires in spaces specialized either physiologically (sexuality) or socially (places set aside, supposedly, for pleasure).
>
> (1974: 391)

Lefebvre is pointing here to the ways in which repressed needs and desires are allowed to be expressed in a socially-sanctioned geography of needs and desires. This argument will be developed in Part III of this book.

16 Such a conceptualisation would avoid making a direct or unmediated link between any social and any personal unconscious; the danger of conflating the social and the personal unconscious is that either social processes are determined by the psychic structures or vice versa.

17 In a Marxist or Habermasian register, we could say that Real spatiality is the site of contradictory organising principles in capitalist social relations, while Symbolic spatiality is constitutive of the legitimation process of contemporary capitalist social relations (see Habermas, 1973). See also Chapter 8 below.

simultaneously overvalues a Lacanian system and thus, unfortunately and dangerously, smuggles in unwanted values: against this, Real, Imaginary and Symbolic spatialities must not be subsumed under a Phallic transcendental signifier, thereby confining each and all to one-dimensional determination. The individual, meanwhile, suffers multiple impulses from these multiple spatialities:[18] the subject must take up a place in relation to these dynamic spatialities, which for the subject represent specific constellations of alienation, power and resistance.

Having talked about Lefebvre's schema for understanding the production of social space in general (and reworked it in Lacanian terms), it is now possible to discuss Lefebvre's three formants of the production of abstract space – in particular, the geometric, the visual and the phallic – to delimit the extent to which a specifically Lacanian psychoanalysis underpins his spatial architectonics and his account of the production of (capitalist) social space.

Mirror images: body, space and the architectonics of the visual

Lefebvre argues that each phase of the history of space is marked by a particular logic of visualisation (to which I will return in the next subsection). Lefebvre's account of visualisation has, however, specific co-ordinates: the mirror, the eye and the gaze, the image. Curiously, these are also Lacan's points of reference. In order to disclose the visual, Lefebvre begins by unpacking the effects of images on space: 'images fragment; they are themselves fragments of space. Cutting things up and rearranging them' (page 97; see also page 313). This analysis is strikingly reminiscent of Lacan's Imaginary, and thus could easily be claimed to delimit Imaginary spatiality. Moreover, in a similar way to Lacan, Lefebvre argues that there is a division between the eye and the gaze; where the eye – or the optical – is associated with the selection of objects and the gaze – or the visual – is associated with detachment. The image, through selection and by detaching the body from space, kills. In a way which parallels Lacan's account of 'lack', Lefebvre argues that the gap which images institute between the body and space has the effect of unleashing desire, a desire which cannot attach itself to any object. Thus, in this Imaginary spatiality, images tend to fragment space but, because these fragments are also abstractions, space cannot provide satisfactory objects of desire. So, for example, the effect on living bodies is to fragment them into particular images, which empty the 'innermost body of all life and desire' (page 98; see also pages 309–310). As for Lacan, both images (as in the mirror phase) and words (as in the entry into language) are

18 This position correlates with Lefebvre's suggestion that the 'I' insinuates itself into the interstices between the real and the imaginary, between bodily space and the body-in-space, (see Lefebvre, 1974: 251).

moments of alienation for the subject.[19]

This account of images relies on Lefebvre's understanding of a particular surface: the mirror – 'in and through the mirror, the traits of other objects in relationship to their spatial environment are brought together; the mirror is an object in space which informs us about space, which speaks of space' (page 186). Like Lacan, Lefebvre is interested in the relationship between the subject, the subject's specular image and the way in which the mirror places the subject in the world. For both, the mirror archetype is also bound up with the separation of the eye and the gaze: the eye that sees the mirror-image is located in the gaze which surrounds both the embodied eye and the mirror; and the eye and the gaze are constitutive of the 'I' (Ego), within the field of the other (see pages 184–186). For both, the mirror is one moment in the alienation of the subject: for Lefebvre (as for Lacan, as in the Narcissus myth), the mirror scotomises ('spirits away') the body, via the 'I' and the eye, through its mirror-image, to the Other (pages 201–202). On the other hand, Lefebvre implicitly (and later explicitly) challenges Lacan's account:[20]

> The interest and importance of the mirror derives not . . . from the fact that it projects the 'subject's' (or Ego's) image back to the 'subject' (or Ego), but rather from the fact that it extends a repetition (symmetry) immanent to the body into space. The Same (Ego) and the Other thus confront each other, as like it as it is possible to imagine, all but identical, yet differing absolutely, for the image has no density, no weight. Right and left are there in the mirror, reversed, and the Ego perceives its double.
>
> (Lefebvre, 1974: 182)

Lefebvre argues that the image and the mirror involve both repetition and difference; thus, the mirror symmetrically duplicates the reflected world, but it also produces a different world in its virtual space. The mirror's reflection produces a doubling of spatiality through repetition and differentiation. This double/d spatiality, which is instituted by the encounter between the body and the mirror, is distinguished by the material qualities of symmetry, images, language, consciousness, time and space. Although the mirror symbolises 'the psychic' in terms of both surface consciousness and deep reflection, this imaginary space and the double/d Ego have material effects. Indeed, these

---

19 As Lefebvre says, 'all of which already seems to suggest a "psychoanalysis of space"' (page 99).

20 Lefebvre also states that:

> The psychoanalysts have made great play with the 'mirror effect' in their attempts to demolish the philosophical 'subject'. Indeed they have gone far too far in this direction, for they consider the mirror effect only out of its properly spatial context, as part of a space internalized in the form of 'topologies' and agencies.
>
> (Lefebvre, 1974: 184, note 18; see also 186)

While there is a point to be made here about the importance of a spatial context, it is not true to say that Lacan ignored spatial context.

effects are so profound that Lefebvre argues that 'it determines the very structure of higher animals' (page 182). Thus, while he stresses the unreality of the specular image, Lefebvre sees the doubling (or mirage) effect of the mirror as constitutive of subjectivity, as does Lacan;[21] unlike Lacan, however, Lefebvre is prepared to argue that the mirror is analogous to social space: that is, that social space acts like a mirror by 'doubling' nature, by mediating between things, by containing and locating bodies and by providing a position, a place in society. For both Lacan and Lefebvre, it seems that the mirror is the archetype for all bodily encounters:

> [Bodies] touch one another, feel, smell and hear one another. Then they contemplate one another with eye and gaze. One truly gets the impression that every shape in space, every spatial plane, constitutes a mirror and produces a mirage effect; that within each body the rest of the world is reflected, and referred back to, in an ever-renewed to-and-fro of reciprocal reflection.
>
> (1974: 183)

In every situation, then, people encounter each other and other objects in an inescapable hall of mirrors, which metonymically and endlessly shuffle part-images between people. The tendency for images to fragment bodies means that

> space is actually experienced, *in its depths*, as duplications, echoes and reverberations, redundancies and doublings-up which engender – and are engendered by the strangest contrasts: face and arse, eye and flesh, viscera and excrement, lips and teeth, orifices and phallus, clenched fists and opened hands – as also clothed *versus* naked, open *versus* closed, obscenity *versus* familiarity and so on.
>
> (1974: 184)

This passage makes Lefebvre's criticisms of Barthes's and Lacan's topologies seem duplicitous (page 5; see above), but the point Lefebvre is making is that the mirror is constitutive of subjectivity by transforming the 'I' into the sign of the 'I': that is, the mirror-image of the body becomes the sign of the body, where the mirror-image is both imaginary and real. The mirror, then, becomes the prototype and paradigm for abstract space: it prioritises the visual (and therefore underpins phallic space), it prioritises the perspectival (and therefore sustains geometric space), it detaches, it alienates and it transforms into images, sign and codes. While Lacan sees the mirror phase as a moment of misrecognition, it is for Lefebvre the archetypal moment of the violence of abstraction. For both, 'the mirror' is simultaneously a trap and a site of alienation: the 'I' must succeed in asserting its difference from its mirror-image or be forever caught like Narcissus (and, for Lefebvre, like

---

21 This can also be seen in Lefebvre's use of the term mimesis (see page 376; for an interesting elaboration of this idea, see Taussig, 1993).

Lewis Carroll's Alice) in the mirror's empty space. Either way, the mirror ensnares the subject in a series of constitutive dualisms, many of which are notably spatial, and which are organised through the material qualities of the encounter between the body and the mirror (page 187):

Symmetry – symmetry/asymmetry
Images – surface/depth, revealed/concealed, opaque/transparent
Language – connoting/connoted
Consciousness – self/other
Time – experienced/repeated
Space – imaginary/real, connected/separated, material/social, produced/producing, immediate/mediated

The mirror naturalises these relationships through repetition and differentiation and the body tends to preserve them. These dualisms also help establish, and maintain, social space: 'Social space can never escape its basic duality, even though triadic determining factors sometimes override and incorporate its binary or dual nature, for the way in which it presents itself and the way in which it is represented are different' (page 191; see also page 411).

Although social space has been characterised as dualistic and abstract space as a mirror, Lefebvre suggests that, while it may speak of Narcissus contemplating himself in the surface of the pool, the mirror metaphor cannot describe the power of landscape to captivate the viewer. The fascinating spectacle of landscape is better thought of, instead, as a mirage. Similarly, Lefebvre is suspicious of the mirror because it cannot elaborate the ways in which a field of vision situates things but does not unify them. A better metaphor, here, would be 'theatrical space' which describes how authors, actors, audiences, characters, scripts and stages are brought together, but never constitute a unity. Indeed, the 'theatre' represents a 'third space' between dualisms such as public/private, seen/unseen, fictitious/real, experienced/perceived (page 188 ff.). Despite this complaint, however, Lefebvre continues to use, and continually returns to, the mirror as the prototype and paradigm of spatial architectonics; in part, this is probably because the mirror is, in Lefebvre's schema, a dialectical unity incorporating image (thesis), mirror-image (antithesis) and field of vision (synthesis). So, even though Lefebvre attempts to use metaphors such as 'mirage' and 'theatre' to suggest that it is inadequate, the 'mirror' remains the underlying and inescapable architecture ('arche-texture') of abstract space and, by extension, social space.

The importance of the mirror is not limited to discussions of 'space'; as for Lacan, the mirror constitutes subjectivity in the encounter with the body. Even while the body produces space, this encounter installs an irrecoverable split (Freud) or gap (Lacan) in subjectivity. That is, for both Lefebvre and Lacan, the mirror subtracts, extracts and abstracts the subject via the eyes, enforcing boundaries between the subject and its others (page 202). The double/d space of the mirror cannot, however, produce one result: while repetition produces the same, differentiation endlessly produces difference; difference which is executed along all the determinants of the field of vision, from the material and dualistic qualities of the mirror, to the third spaces beyond the mirror (see pages 188 and 195).[22] The mirror, then, mobilises a sense of space as fixed and fluid: fixed by repetition, made fluid by difference. There are anchoring points in this architecture – and these are sites of power, marked by a particular scar: the phallus:

> The arrogant verticality of skyscrapers, and especially of public and state buildings, introduces a phallic or more precisely a phallocratic element into the visual realm; the purpose of this display, of this need to impress, is to convey an impression of authority to each spectator. Verticality and great height have ever been the spatial expression of potentially violent power.
>
> (1974: 98)

There is a link in Lefebvre's work, as in Lacan's, between the visual, the phallic and regimes of power.

Phallic space(s): abstraction, power and the logic of the visual

The ravages of the mirror – as prototype and paradigm of image-production – may produce double/d bodies in parts: but the Imaginary spatiality of the body is marked by two specific organs – 'the eye' and 'the phallus' (1974: 204; see also 302). As for Lacan, the phallus is the eye in the storm of meaning; the phallus is the signifier that simultaneously veils, guards and represents power (see, for example, page 408). The Phallus then is a signifier, erected both on the violent fragmentation of the body and on the body's abstraction via the eyes, which simultaneously highlights and hides (deathly) power.[23] The phallus, thus, becomes central to Lefebvre's argument that the social space of capitalist social relations is characterised by the violence of abstraction (see page 289; similarly page 350). This abstract space

> endeavours to mould the spaces it dominates (i.e. peripheral spaces), and it seeks, often by violent means, to reduce the obstacles and resistance it encounters there ... A symbolism derived from the mis-taking of sensory, sensual and sexual which is intrinsic to the things/

---

22 Two examples of a third space might be 'the unconscious' and 'language', both of which lie beyond and between the body/Ego dualism, see pages 203 and 208.

23 As Lefebvre argues,

> Monumentality, for instance, always embodies and imposes a clearly intelligible message. It says what it wishes to say – yet it hides a good deal more: being political, military, and ultimately fascist in character, monumental buildings mask the will to power and the arbitrariness of power beneath signs and surfaces which claim to express collective will and collective thought.
>
> (1974: 143)

> signs of abstract space finds objective expression in derivative ways: monuments have a phallic aspect, towers exude arrogance, and the bureaucratic and political authoritarianism immanent to a repressive space is everywhere [see also pages 142–143] ... A characteristic contradiction of abstract space consists in the fact that, although it denies the sensual and the sexual, its only immediate point of reference is genitality: the family unit, the type of dwelling ... fatherhood and motherhood, and the assumption that fertility and fulfilment are identical.
>
> (1974: 49–50)

The abstraction by the visual leads to the failure-to-recognise what Lefebvre calls the total body; instead, certain body parts are raised up and brutally prioritised over others. In particular, the penis is symbolised in the form of monumentality. This form is not of course unique to either abstract space or capitalist social relations, nevertheless phallic verticality reaches new heights under these conditions (see pages 144–147). An irony of this situation is that the increasing elevation of the phallus, as symbol of the penis and rampant (male) sexuality, is associated with both genitality (following, it seems, Freud's characterisation of adult sexuality as genital-centred) and the conflation of procreation and pleasure in the ideology of the family. Lefebvre is not simply pointing to the organisation of capitalist social relations in the abstract or the concrete form of the family, he is also alluding to the organisation of space: for example, noting that major Paris streets have monuments at the points where they intersect and that house design assumes a particular familial ideal (see Chapter 7 below).

The problem is that this spatial organisation gives the impression of intelligibility; it appears to be the expression of a collective, shared project – whether monuments to heroes and victories, urban planning or home building. These forms, however, hide what is at issue: power. The control of space presumes, establishes and maintains an arrogant will to power and a phallic space is produced which acts to reproduce the value systems of the powerful. Space is marked, then, by the phallus, arrogance, the will to power and masculine brutality. It is a repressive space which does not allow or permit other spaces, such as female and/or feminine spaces (pages 262 and 379–380).

Once more, psychoanalytic themes – repression, the phallus – are tied to an analysis of the abstract, that is, of the production of space under capitalist social relations. For Lefebvre, it seems, capitalist spatial relations cannot be understood without a psychoanalytic imagination. Further, as I have shown, Lefebvre suggests that abstract space transports the body outside itself into a visual regime and that this has serious consequences: the body is fragmented into images, pleasure is foreclosed, the reciprocity (Lefebvre presumes) of the sexual relationship is undone. Cold abstraction produces a spatiality without pleasure, a castrated space – real and imaginary, concrete and symbolic. This undoubtedly psychoanalytic explication of abstract capitalist socio-spatial relations is further prosecuted; Lefebvre continues in a distinctly Lacanian mood:

> the space of a metaphorization whereby the image of the woman supplants the woman herself, whereby her body is fragmented, desire shattered, and life explodes into a thousand pieces. Our abstract space reigns phallic solitude and the self-destruction of desire. The representation of sex thus takes the place of sex itself, while the apologetic term 'sexuality' serves to cover up this mechanism of devaluation.
>
> (1974: 309)

Here, Lefebvre is undeniably drawing on masculinist assumptions when he cuts up the woman's body and forces it to bear the burden of 'sex' and 'sexuality' (pages 309–310). Moreover, his assumption that there was a body which had meaning in and of itself before its fragmentation and abstraction is also highly dubious. It seems clear, though, that Lefebvre's analysis directly correlates with Lacan's equally doubtful understanding of the phallus and femininity. Thus, abstract space is organised through castration, veiled and revealed in the phallus. The image of woman replaces the woman (through metonymic substitution). This, of course, accords with Lacan's suggestion that 'there is no such thing as *The* woman' and that 'Woman' is a signifier that derives its meaning only in relation to the phallus (see Lacan, 1972–3: 144). Indeed, Lefebvre, like Lacan, identifies the penis and the phallus. On the other hand, there is an attempt in Lefebvre's work (as in Lacan's) to separate phallus (signifier) and signified (penis), which at least offers the possibility of non-biologically determinist understandings of the relationships between male bodies, masculinist representations and power (see Lefebvre's criticisms of the conflation of 'speech' and 'penis' in note 27: 262). Nevertheless, the assumption of the centrality of the phallus in the realm of speech, the failure to specify the relationship between the phallus (speech) and the penis (body) and the unquestioning repetition of the denigration of female genitality lead Lefebvre to draw on familiar masculinist understandings of phallic power:

> The phallus is seen. The female genital organ, representing the world, remains hidden. The prestigious Phallus, symbol of power and fecundity, forces its way into view by becoming erect. In the space to come, where the eye would usurp so many privileges, it would fall to the Phallus to receive or produce them. The eye in question would be that of God, that of the Father or that of the Leader.
>
> (1974: 262)

Although there are masculinist presumptions here – for example, woman-as-world, vagina-as-hidden, Phallus-as-prestigious – there is also an explicit critique of masculine space, of space ruled by the deathly phallus. There is also an implicit reference to Freud's assessment of the church and group leader (see Chapter 4 above), but this is situated within a story about 'the eye'. For Lefebvre, the visual enables abstraction. It also organises and polices space: 'Abstract space is doubly castrating: it isolates the phallus, projecting it into a realm outside the body, then fixes it in space (verticality) and brings it under the surveillance of the eye ... because of the process of localization, because of the fragmentation and specialization of space' (page 310).

Abstract space is a site of alienation, separating the 'I' from its embeddedness in the material world, by placing images and signs within a particular logic of the visual: the eye is dissociated from the visual field and thus space only appears in reduced forms; the eye sees what it needs to see, ignoring everything else; and the eye sees in the way it needs to see, failing to look in other ways. Abstract space 'offers itself like a mirror to the thinking "subject", but, after the manner of Lewis Carroll, the "subject" passes through the looking-glass and becomes a lived abstraction' (pages 313–314). The 'subject' is abstracted by the logic of visualisation. For Lefebvre, the logic of visualisation incorporates elements of the visual, power and space; in this respect he suggests that abstract space is made up of the three formants of geometric, optic and phallic space (pages 285–287). There are two aspects of this spatio-visual regime and these correlate with two figures of speech – metaphor and metonymy (see especially pages 98–99 and 286–287): that is, a metaphorical logic of spatialisation, which refers to the ways in which bodies are caught in webs of analogons such as images, signs and symbols; and a metonymic logic of spatialisation, which is about 'a continual to-and-fro ... between the part and the whole' (page 98).

Through the optical formant of abstract space, the visual has come to dominate the whole of social practice (according to Lefebvre, page 286). Within a psychoanalytic perspective, abstract space is dominated by the phallogocentric organisation of the visual and the symbolic. However, drawing as much on McLuhan and Debord as *The Interpretation of Dreams*, Lefebvre suggests that there are two sides to abstract socio-spatial practices: the written word, which is metaphorical, and the spectacle, which is metonymic. The word and the spectacle have become predominant, at the expense of all other bodily senses.[24] By alienation and abstraction (through the eyes), the word and the spectacle act to produce a space empty of meaning. Meanwhile the phallic formant, which papers over this gap with 'full' objects, metaphorically symbolises power and metonymically represents specifically masculine brutality. There is a Lacanian story here: there is a 'gap' opened up by the eye and by space (the mirror, the geometric), which is belied by a language organised around a veiling and revealing signifier (the phallus). As in Lacan, however, this situation remains unexplained. Nevertheless, Lefebvre's commitment to appropriating and relocating psychoanalytic terms of reference is instructive.

In sum, Lefebvre's appropriation of psychoanalysis is woven into his account of the production of space at strategic moments: 'social space', 'abstract space' and 'spatial architectonics' cannot be understood without a

---

24 There is of course a long history to this perspective, see for example Gay, 1984 and Jay, 1993.

psychoanalytic appreciation of Lefebvre's deployment of notions of desire, pleasure, phallus, mirror, Ego, prohibition, castration and so on. It is possible, to summarise very quickly, to take away from this juxtaposition a sense that 'the mirror' is an important metaphor for spatial relationships; although, in the spirit of Lefebvre, this has to be further situated along the lines I suggested at the end of Chapter 5. Similarly, 'the phallus' needs to be torn down from its lofty place. Nevertheless, Freud, Lefebvre and Lacan suggest that an account of space must be watchful for specific spatial relationships:

1. the relationship between 'objects' (understood psychoanalytically),
2. the repeated hiding of things (such as trauma, or power) in the open, though in disguised forms (such as the symptom, or monument),
3. living in a hall of mirrors, in shadows, echoes and mirages,
4. the self presentation of the subject as a masquerade which defines the subject,
5. real, imaginary and symbolic spatialities, and
6. the placing of the subject within multiple, interacting relations of alienation, power and resistance.

In some ways, however, these six points represent questions which remain to be solved. They also refer back to some psychoanalytic concerns which have been passed over, or assumed away, in the work of Freud, Lacan and Lefebvre. An account of space must take these questions about the relationship between the body and space, subjectivity and space, society and space further; but in directions which do not melancholically presume some ideal past, or presume that there is nowhere to go, or offer false promises, or victimise people – and which allow for the possibility that people make history, though not in circumstances of their own choosing.

Conclusion: mourning for Lefebvre and psychoanalysis

Both Lefebvre and Lacan mourn for something (a body, a space) which has been lost. Lefebvre mourns for some primeval moment when knowledge and social practice were inseparable, a time when bodies and nature were inseparable. While Lacan mourns for the time before lack constituted the subject, Lefebvre seems to suggest that this lack was itself instituted by the dissolution of his grieved-for, absolute space (see page 242). In their melancholia, they offer similar appraisals for the prospects of recovering from this disaster. In a moment which is surprisingly evocative of Lacan's morbid sense that it is already too late for us, Lefebvre argues, 'the fact remains that it is too late for destroying codes in the name of a critical theory; our task, rather, is to describe their already completed destruction, to measure its effects, and (perhaps) to construct a new code by means of theoretical "supercoding"' (page 26).

Perhaps their melancholy and their reliance on 'supercoding' can itself be placed alongside other psychoanalytic discourses, which are not quite so ready to think only after the already completed destruction and which accept the full play of presence and absence.[25] Indeed, the idea that 'we' all *had* 'it' may be seen to be part of a masculine imaginary and the idea of recovering 'it' through supercoding may be seen as a specifically masculinist response to the problem. There are other possibilities.

CONCLUSION: THE PSYCHOANALYSIS OF SPACE

In the previous section, I suggested that a psychoanalytically-informed account of space would draw on several resources:

1 a psychoanalytic understanding of the relationship between the body, the psyche and the subject's place in the world;
2 a sense that subjectivity is played out through fixity and fluidity;
3 an understanding of the subjects, objects and spaces which relies on their partiality, their duplicity, their trickery and their (supposed) truth;
4 an awareness that subjects act out the truth of their subjectivity as a situated and repeated performance, but a performance which is itself the 'truth' of subjectivity;
5 a differentiation between dissimilar (psychic, social) spatialities and the often conflictual relationships within and between them; and
6 an alertness to social sanction, social power and the possibilities of resistance – indeed, it ought to be quickly added that if psychoanalysis does not inform a politics of resistance, of position and of subjectivity, then there is little point in using it.

Let me elaborate, very briefly, on each of these points. I will signpost the work of other writers to suggest ways in which masculinist presumptions found in Freud, Lacan and Lefebvre may be put to sleep.

Freud offers distinct, dynamic models of the mind, such as id, ego, super-ego and unconscious, preconscious, conscious. These models undo static notions of the mind. They also disrupt any sense that individuals are autonomous and/or that they are separable from their circumstances; on the other hand, Freud's topologies of the mind also locate the individual within their circumstances – psychic and social (to pull out only two co-ordinates). His analysis circulates around the child's body, the child's understanding of its body and the socially-sanctioned meanings that enmesh the child; this can be seen in Freud's appropriation of the Narcissus and Oedipus myths. He proposes various mechanisms through which the child's emplacement and displacement take place: identification, idealisation (to which abjection may be added, following Kristeva, 1980), incorporation, introjection (and projection) and internalisation (see Chapter 4 above). The example of the Fort/Da

25 Please refer to the discussions of the Fort/Da game in Chapters 5 and 7 of this book. See also Freud, 1920: 283–286, for a sense of this 'full play'.

game shows that each of these 'placings' is actively constituted by space (see also Chapters 5 and 7): it matters that the cotton reel is 'there', then 'here'; indeed, it is impossible to imagine subjects, subjectivity and the relationships between subjects except spatially; even so, distinct spatialities are being evoked and different, shifting boundaries drawn. Following a favourite metaphor of the Frankfurt School, it might help to think of a 'constellation' of relationships; where the metaphor-concept constellation highlights aspects of looking, resolution, the organisation of the seen and the unseen, depth and myth.

Another concern of psychoanalysis and, to a much lesser extent, of 'spatial' analysis is the sexed body;[26] especially the valuation of specific body parts (such as the penis) as opposed to others (the vagina). The position of the female – and especially the mother's – body becomes central to this analysis, as shown by the work of Horney (1932, 1933). Horney has been criticised for her tendency to rely on biological difference as an explanation of the difference between men and women; on the other hand, her work is exceptionally interesting as a description of the ways in which women's bodies become a site of terror and horror – for both men *and* women. This observation is crucial: if only men are seen as denigrating women's bodies and are thereby named as the 'agents' who produce masculinism, then women become either passive victims or unknowing dupes of male-organised and masculine-centred values. Horney's work suggests that women too have their (different) investments in, and (different) collusions with, masculinist maps of meaning, identity and power. Moreover, this critique of fantasy, ambivalence and collusion can be extended to other geographies of meaning, identity and power and to other 'fragments' of the body.[27]

Embodied subjects place themselves into topographies of meaning, identity and power which value certain aspects of bodies and subjectivity more highly than others: people are expected to be and behave along lines sanctioned by society. As the eye of the subject sees, so it is also subjected to the gaze of the visual regime of others. The (Lacan's *and* Lefebvre's) differentiation of the eye ('I') and the gaze (seeing others)[28] suggests a subject who owes obedience to the law (the 'thing' under discussion) but also creates meaning and has power (the 'person' who brings things under their control). These ambiguously-placed, embodied subjects, in psychoanalytic terms at least, never fit the bill: they must repress unacceptable or dangerous impulses if they are to become 'civilised' – like the ugly sister in the fairy tale who cuts her toes off to fit into Cinderella's shoe.

Space intervenes to cut up and re-present the embodied subject in specific

26 See, for example, essays collected in Parts II and IV of Pile and Thrift, 1995a.

27 These geographies can be found in the work of Fanon, 1952, Theweleit, 1977, 1978 and Irigaray, 1974, 1977. These arguments are followed up in Chapter 6 and the Conclusion to this book.

28 Others who see and others who are seen.

ways in particular circumstances; and, on the other hand, these subjects learn to read their selves off from the reflections they see in the hall of mirrors in which they are placed, producing 'identities' which are best understood as (their) masquerades (see Riviere, 1929, on this). The sense that identity is a masquerade subverts understandings of subjectivity which are grounded either/both in essentialisms or/and in dichotomies of identity and power, such as masculine/feminine, black/white, male/female, heterosexual/ homosexual. On the other hand, 'masquerade' does not reduce subjectivity to a free-floating whirl of signifiers: thus, while the subject is placed within matrices of meaning, identity and power, these are actively produced and regulated as 'practices of gender coherence' (Butler, 1990: 24). Here, psychoanalytic readings of the embodied subject work through the 'fixed' and 'fluid' points of capture of subjectivity, space and power.

Finally, psychoanalysis highlights the positions, directions, distances, shifts, folds and tears (and tears) of subjectivity. Subjectivity and space are interrelated; power infuses subjectivity and space; and power relationships are multiple and interacting, have many layers and different dynamics (and so Lefebvre's and others' privileging of 'production' becomes wholly inadequate). In this 'interanimation', space is not a backdrop – a passive, fixed arena in which 'stuff' takes place. Constellations of relationships in Freud's myths – Oedipus etc. – are drawn out of spaces and the gaps between. Thus, the Narcissus story is incomprehensible without understanding that the 'confusion' of the mirroring pool is its constitutive 'virtual' and 'real' spatialities. Spatiality enables the repression of unacceptable impulses; it allows loss to be made up or papered over; it permits 'symptoms' to be written out; and so on. Spatiality constitutes the performance of individuals in different social spaces (to appropriate arguments by Goffman, 1959); where this spatiality shifts around real, imaginary and symbolic co-ordinates, but where the 'content' of the real, the imaginary and the symbolic cannot be presumed in advance. Spatiality enables power to be exercised; in this sense, it is a modality through which power relations are normalised, naturalised and neutralised.[29]

So far, it has been suggested that subjectivity, space and the social are simultaneously fixed and fluid; that an understanding of this requires the use of different conceptual maps; that these maps should outline the struggle for coherence as much as the points where that coherence breaks down; and that these maps are political. It is now possible to see where these cartographies might take us.

29 For a brief elaboration of this argument, see Chapter 12 of Keith and Pile, 1993b.

Part III

THE SUBJECT OF SPACE

INTRODUCTION TO PART III

In his analysis of the production of space, Lefebvre makes a distinction between spatial practice, the representation of spaces and representational spaces (1974: 33; see also the Conclusion to Part II above). These characterise social space or aspects of social space. Interestingly, Lefebvre's choice of example, when he elaborates the relationship between them, is the body:

> social practice presupposes the use of the body ... This is the realm of the *perceived* (the practical basis of the perception of the outside world, to put it in psychology's terms). As for *representations of the body*, they derive from accumulated scientific knowledge ... Bodily *lived* experience, for its part, may be both highly complex and quite peculiar, because 'culture' intervenes here, with its illusory immediacy, via symbolisms and via the long Judaeo-Christian tradition, certain aspects of which are uncovered by psychoanalysis. The 'heart' as *lived* is strangely different from the heart as *thought* and *perceived*.
>
> (Lefebvre, 1974: 40)

Through the body, the three aspects of social space have been translated into a different register: the perceived, the conceived and the lived. Broadly, by locating his schema in the use of the body, Lefebvre has converted spatial practice into the realm of the perceived, representations of space into the conceived and representational space into the lived. It would appear that understanding social space in the abstract requires some grounding in the bodily experience of everyday life. More than this, it seems that producing this embodied understanding of social space requires a change in the frames of reference within which social space is understood in abstract terms. This enhanced map of bodily social space has psychoanalytic referents: Lefebvre argues that social prohibitions and the tendency for language to fragment the body produces a moral geography of the body which is marked by the removal of its organs and by the erasure of its needs and desires.

On the other hand, I have shown that Lefebvre's argument that psychoanalysis is only able to elucidate 'lived' or 'representational' aspects of social space fails to appreciate not only the extent to which psychoanalysis might contribute to an understanding of all aspects of social space but also the ways in which psychoanalysis might extend an understanding of space. Lefebvre's choice of the body is, however, instructive: the body would be one place from which to develop a psychoanalytically-informed discussion of space. A second site is provided by Lefebvre as he elaborates the perceived, conceived and lived spaces of the body. He begins by arguing

> that the lived, conceived and perceived realms should be interconnected, so that the 'subject', the individual member of a given social group, may move from one to another without confusion – so much is a logical necessity. Whether they constitute a coherent whole is another matter. They probably do so only in favourable circumstances, when a common language, a consensus and a code can be established.
>
> (1974: 40)

Lefebvre suggests that this common language, which coheres the bodied subject and the social, the lived, conceived and perceived without confusion, actually existed in a particular place: cities in Renaissance Italy. This situation existed both because representations of space (the conceived) tended to dominate representational spaces (the lived) and because this circumstance was bound up with a particular logic of visualisation grounded in perspectival space, which is characterised by 'a fixed observer, an immobile perceptual field, a stable world vision' (page 361). The urban effect in Renaissance Italy is that people lost their 'place in the "world" and Cosmos' and found themselves instead in the city (page 272). The Renaissance city developed along the lines and co-ordinates set by a (perspectival, linear, geometric) conception of space which was embedded in political power, where political power was not only reproduced and maintained through this social space but also did so by excluding other (for example, cosmological) concepts of space.

There is, here, some resonance between the spaces of the city and the space of the body which has something to do with the way in which (in this example) political power is mobilised, thus: *either* both the spaces of the body that sees and recognises and of the city that is conceptualised visually and spatially were brought into alignment through political power; *or* their alignment (for other reasons, such as advances in science) enabled a particular kind of political power to be effected. I believe that both of these possibilities can be understood psychoanalytically – and, moreover, that this 'psychoanalysis of space' might help untangle the relationship between subjectivity, spatiality and power relations.

If it is interesting that Lefebvre chooses 'the body' and 'the city' to exemplify the production of space, then it is remarkable that he utilises psychoanalysis to explain these moments of social space. It is in these places, then, that it is possible to look for psychoanalyses of space. Of course, bodies and cities have intricate interwoven histories (see Sennett, 1994). Moreover, it is in many ways artificial to eviscerate bodies from cities. So, before outlining the argument and content of Chapters 6 and 7, I would like to preface them with an illustration of the ways bodies and cities 'produce' one another; hence, in the next subsection, I will discuss selected aspects of Peter

Stallybrass and Allon White's *The Politics and Poetics of Transgression* (1986).

THE SEEN AND THE SCENE: THE BODY AND THE CITY

Stallybrass and White's book, *The Politics and Poetics of Transgression*, is organised through many themes which lie at the centre of my concerns – not least psychoanalysis, space, bodies, cities – and their framework for looking at the relationship between these 'things' is also inspiring, subtle and insightful. Their work involves disclosing the ways in which European cultures are ordered and maintained through the social distinction between 'high' and 'low' value. They argue that 'maps of meaning, identity and power' are produced through co-ordinates which are placed on opposing poles;[1] these poles are then associated either with a high valuation and associated with the powerful, or with low valuation and associated with the powerless.

Stallybrass and White argue that there are four fundamental symbolic domains in which these valuations take on special significance for the ordering and maintenance of social distinctions: these are psychic forms, the body, geographical space and the social order. Although their empirical concern is with literary genres, Stallybrass and White see these as 'a particularly clear example of a much broader and more complex cultural process whereby the human body, psychic forms, geographical space and the social formation are all constructed within interrelating and dependent hierarchies of high and low' (1986: 2). They are not content, however, to reproduce – and therefore collude with – those maps of meaning, identity and power; their analysis is also intended to uncover transgressive elements which either invert categories of high and low or slip between them.

There is slippage between high and low valuations partly because there is an internal relationship between them: i.e. high is not-low and low is not-high, the one is defined in opposition to the other and makes no sense outside of this opposition. From this perspective, while the powerful are busy proclaiming themselves as being of high value, they can only do so by finding, naming, ordering and maintaining others who are not of high value. The self-proclaiming high is constituted by an other who, though necessary to high self-identity, is denigrated and seen as both opposite and separate: such that

> the top *includes* that low symbolically, as a primary eroticized constituent of its own fantasy life. The result is a mobile, conflictual

1 The phrase 'bodies are maps of power and identity' is Donna Haraway's (1990: 222). As you may have noticed, I like turning this idea around in different contexts to extend and elaborate its meaning (and power).

> fusion of power, fear and desire in the constitution of subjectivity: a psychological dependence upon precisely those Others which are being rigorously opposed and excluded at the social level. It is for this reason that what is *socially* peripheral is so frequently *symbolically* central.
>
> (1986: 5)

Stallybrass and White are drawing on a psychoanalytic understanding of the constitution of subjectivity to talk about the ways in which the self is oriented around bi-polar axes: 'I am like that' and 'I am not like that' with 'I want to be like that' and 'I don't want to be like that', and so on. In particular, these bi-polar oppositions accord with psychoanalytic notions of abjection, which have been elaborated by Kristeva in particular (1977, 1980; see also Fletcher and Benjamin, 1990). It is from this perspective that Stallybrass and White suggest that there is an internal relationship between power, desire and disgust (an idea which recurs throughout their work), which is not only conflictual and fluid but also fused and fixed.

Maps of meaning, identity and power are charted by identifying what is high and not-high and by establishing differences between the high and the low, which are defended by keeping the low–Others at a distance. However, the identification and defence of high/low boundaries can only be achieved through constant vigilance: the powerful are constantly looking with desire and disgust, in fascination, at things which are considered outside their selves. This can be illustrated by Stallybrass and White's discussion of dominant constructions of the body in European cultures. They argue that there is a categorical distinction between a 'high' body and 'low' representations of the body which they term classical and grotesque:[2]

> The classical statue has no openings or orifices whereas grotesque costume and masks emphasize the gaping mouth, the protuberant belly and buttocks, the feet and genitals. In this way the grotesque body stands in opposition to the bourgeois individualist conception of the body, which finds *its* image and legitimation in the classical. The grotesque body is emphasized as a mobile, split, multiple self, a subject of pleasure in processes of exchange; and it is never closed off from either its social or ecosystemic context. The classical body on the other hand keeps its distance. In a sense it is disembodied, for it appears indifferent to a body which is 'beautiful', but which is taken for granted.
>
> (1986: 22)

The distinction between the (grotesque, ugly) body-in-place and the (classical, beautiful) body-out-of-place is important because these representations 'speak' social, spatial and psychic relationships between differently-

---

2 Lynda Nead draws a similar conclusion in her analysis of 'naked' versus 'nude' bodies in western art, 1992; see also Russo, 1994.

located people – whether by class, race, gender, sexuality and so on.[3] The point they are making is that these grids of meaning, identity and power are impossible without distinctions between high and low, without these maps being 'spoken' in different symbolic domains such as the body and geographic space, and without these maps being articulated through geographic spaces like the body and the city.

For Stallybrass and White, subjectivity is formed through chains of associations ('transcodings') where one 'high' value is associated with another 'high' value produced in a different discursive domain and where this is then associated with others. These associations are made primarily through figures of speech, such as metaphor and metonymy (and this links to psychoanalytic understandings of the relationship between subjectivity, meaning and power: i.e. these associations are neither 'free' nor transparently obvious). The interactions between these associations produce intricate and dense matrices of meaning which are 'topographical' in the sense not only that difference is produced 'spatially' but also that matrices are played out in specific sites. Thus, for example, 'discourses about the body have a privileged role, for transcodings between different levels and sectors of social and psychic reality are effected through the intensifying grid of the body' (page 26).

If this statement was rewritten by replacing 'the body' with 'the city', it would not damage Stallybrass and White's point or change their argument: that is, both the body and the city are intensifying grids for simultaneously social and psychic meanings, produced in the mobile, conflictual fusion of power, desire and disgust. In this spirit they end their chapter on the city by concluding that: 'the reformation of the senses *produced*, as a necessary corollary, new thresholds of shame, embarrassment and disgust. And in the nineteenth century, those thresholds were articulated above all through specific *contents* – the slum, the sewer, the nomad, the savage, the rat – which, in turn, remapped the body' (page 148).

They conclude that there is an internal relationship between power, fear (expressed as shame, embarrassment and disgust) and desire, which is recursively mapped onto bodies and cities through specific co-ordinates such as the sewer,[4] the slum, the rat, the nomad,[5] the analysand, the proletarian,

---

3 For a discussion of the ways in which these high/low valuations of the body are performed and transgressed in Rio Carnival, see Lewis and Pile, 1996.

4 Lefebvre observes that the *mundus*, the pit in the middle of mediaeval towns where rubbish was thrown, was associated with particular parts of the body (1974: 242). It was not just seen as an anus, but also as a passageway and thereby associated with both birth and tombs (death), and as a mouth into earth-as-mother it was also associated with the vagina. This example shows the ways in which chains of associations move from cityspace to bodyspace by transcoding elements related to each other by social and spatial valuations (as Stallybrass and White argue). Immediately after this example, Lefebvre talks about a psychoanalysis of space and the psychic realities of space. Later, Lefebvre pushes this analysis further by arguing not only that the Ego takes refuge in the pit, but also that real and imaginary spaces like the *mundus* exist in 'the unassignable interstice between bodily space and bodies-in-space (the forbidden)' (1974: 251). This space might now be understood to be a 'third space'.

5 Stallybrass and White's account of fears of the nomad who transgresses social and spatial boundaries has contemporary resonance with Sibley's uncovering of the psychodynamics of

the bourgeois and the prostitute. Each location 'speaks' at once social and psychic, simultaneously real, imaginary and symbolic categories of fear and disgust. Indeed, in support of this assertion, it is possible to cite a personal anecdote which Freud uses to describe the disturbance created by uncanny repetition (1919: 359). He tells a story about wandering around a small provincial Italian town, when he inadvertently ends up in the red light quarter. He tries to leave this street only to return to it not once but twice. Freud's ambivalence is marked by a sense of uncanniness, excitement and discovery. In terms of these sites of power, disgust and desire, Stallybrass and White immediately continue,

> it is important to emphasize that this 'manifest content' was no incidental and contingent metaphor in the structuring of the bourgeois Imaginary. It was not a secondary over-coding of some anterior and subjective psychic content. Indeed it participated in the constitution of the subject, precisely to the degree that identity is discursively produced from the moment of entry into language by ... oppositions and differences.
>
> (1986: 148)

Reading the city and the body, for Stallybrass and White, requires an analysis of their 'manifest contents', how these relate to the structure of specific Imaginaries and how these Imaginaries participate in the constitution of subjects located in language through oppositions and differences. Such a reading is psychoanalytic: 'manifest content' and 'secondary revision' are ideas used by Freud in the interpretation of dreams,[6] while 'Imaginary' and 'entry into language' are derived from a Lacanian lexicon (see Chapter 5 above). The point they are making is that 'contents' such as 'the prostitute' and 'the slum' or 'the bourgeois' and 'the suburb' are constituted by the internal relationship between power, desire and disgust and are not the expression of some a priori autonomous essence. Through the bi-polar encoding of specific Imaginaries, these sites become intensifying grids for the reproduction of simultaneously social and psychic categories of distinction (between 'high' and 'low'). The body and the city are cartographies of meaning and identity; they are intensifying grids of power, desire and disgust. Thus, they begin their analysis of the city in this way:

---

the social and spatial exclusions experienced by gypsies (see Sibley, 1992, 1995b). See also the Introduction to Part II above.

6 'Manifest content' refers to the dream as it is dreamt by the dreamer, and has been taken to refer to any representation (e.g. the word as it is spoken by the speaker, the picture as it is seen by the viewer, and so on) before it is 'analysed', whether by psychoanalysts or not. Manifest content is counter-posed to 'latent content' which is the meaning of the dream in terms of its unconscious language. 'Secondary revision' refers to the revision of the dream as it is recounted by the dreamer. These terms appear in *The Interpretation of Dreams* (Freud, 1900), but they have gained some purchase in psychoanalytically-inspired literary theory (see Wright, 1984).

> In the nineteenth century [the] fear of differences that 'have no law, no meaning, and no end' was articulated above all through the 'body' of the city: through the separations and interpenetrations of the suburb and the slum, of grand buildings and the sewer, of the respectable classes and the lumpenproletariat.
>
> (1986: 125)

A bourgeois Imaginary saw the 'lower' classes as ignoring the moral codes necessary for respectability: this 'moral laxity' produced an ambivalent gaze, because moral looseness was simultaneously threatening and absorbing. Thus, low–Others were seen to be dirty, diseased, criminal and sexually promiscuous; on the other hand, such 'freedom' from moral restriction was fascinating (both captivating and captive-making); the bourgeois observer could hardly keep their eyes off such behaviour. This fascination (desire and disgust) with low–Others was mapped across the topography of the city – from the suburb to the slum, from the bourgeois reformer to the prostitute.

Stallybrass and White argue that new forms of social regulation produced a particular urban geography which was constructed out of this bourgeois Imaginary (page 126; see also Elias, 1939 or Gay, 1984). However, the boundaries which parcelled up this urban geography were continually transgressed: the sewer ran between the suburb and the slum bringing with it, according to commentators, the possibility of contamination and pollution; while the prostitute came from the slum to walk the streets suggesting, for contemporaries, both the possibilities of 'pleasure' and other kinds of 'pollution'. It is worth recounting this analysis of the sewer in a little detail because it sketches themes which will be reviewed in greater detail in Chapters 6 and 7 of this book.

In the nineteenth century city, the bourgeois gaze was confronted by a spectacle of poverty: the primary focus of this observation was 'cleanliness'; that is, observers were preoccupied by the mess and filth of low–Others,[7] not least because, in their fears, 'dirt' and 'smell' transgressed the boundaries between the clean suburbs and the filthy slums. Thus, 'dirt' became part of a discourse about contamination whether by sight, touch or smell (pages 130–131 and 139–140),[8] and the 'Dirt' of these low–Others was increasingly emphasised in reforming discourses. The regulated body was charted in the regulated city: clean bodies needed a proper sewage system.

---

7 These low–Others included (according to Mayhew and Marx) vagabonds, beggars, scavengers, rag-gatherers, pure-finders (pure meant shit), bone-grubbers, sewer-hunters, mudlarks, brothel keepers, prostitutes, tinkers, knife-grinders, rat-killers, organ-grinders, crossing-sweepers, dustmen, gamblers, pickpockets, thieves, discharged jailbirds, swindlers and cheats (Stallybrass and White, 1986: 129–130).

8 Stallybrass and White's stress on sight, touch and smell is reminiscent of Lefebvre's insistence that the (total) body should not be reduced to one sense: sight. These observations about 'dirt' resonate strongly with Mary Douglas's classic study of pollution and taboos (1966). Remember also the work of David Sibley, described in the Introduction to Part II above.

Another aspect of the relationship between the slum and the bourgeois gaze is revealed by Engels. He described a situation in Manchester where the slums were hidden behind the respectable façade of the streets, which were lined with shops (Engels, 1844). In this way, the 'dirty' were rendered invisible to the bourgeois by the urban geography of the nineteenth century city. Moreover, as part of this critique, Engels also traced the sewer back to the bourgeois and traced the transgressions of body 'dirt' across a socially and spatially divided city. Nevertheless, Engels – as Stallybrass and White point out – had his own demon others: the Irish, whom he likened to pigs. Thus Engels relied on a colonial Imaginary whose desire and disgust concerned the 'savage' and 'dirty' lives of other 'races' (see Stallybrass and White, 1986: 130–131). This Imaginary segregated out low–Other races, and sometimes it was organised through mobilising metaphors such 'the pig'. Thus, dirty conditions were associated with certain peoples, who must live like pigs and who were thus condemned as being pigs.

Similar metonymic chains of associations were built up around other low–Others: thus, slums were linked to dirt, dirt to sewage, sewage to disease, disease to moral degradation, and moral degradation to the slum-dweller or to the prostitute. Through metaphorical and metonymic association, the slum-dweller was a pig who lived in filthy slums of their own making. Sliding signifiers are elided; metaphor and metonymy enable the condensation and displacement of one expression with another; the entry into language is complicit in the social and psychic construction of an embodied urban geography mapped and remapped through power, disgust and desire.

The surveillance and policing of slums could not stop at their 'physical' limits: dirty bodies could not be confined to this filthy space. As the bourgeoisie saw it, the problem of contamination was bound up with the 'movement' of people through the city. The city was a site and sight of promiscuity: people 'mixed' in public spaces. The 'refined', 'civilised' and 'healthy' bourgeois felt under siege by 'crude', 'savage' and 'diseased' Others, who they felt to be spiritually inferior: crowds were repugnant. Moreover, in the streets, the bourgeois gaze might be confronted by the other who did not look back with deference, respect or passivity; the bourgeois might even be 'humiliated' by the touch of a low–Other. Yet the spectacle of the city fascinated the bourgeois, while the prostitute promised another kind of contact: 'It was above all around the figure of the prostitute that the gaze and the touch, the desires and contaminations, of the bourgeois male were articulated' (1986: 137).

The bourgeois Imaginary was not just organised around class (the bourgeois over and above the proletariat and lumpenproletariat) and racialised (the bourgeois white over and above the 'savage', the 'primitive', the 'swarthy' Irish, and so on), it was also gendered male: the figure of the prostitute carried contradictory and ambivalent male desires and fears. The female body of the prostitute became an intensifying grid in a discourse about contagion, pleasure and sin. In bourgeois fantasies, she moved between the dirty and filthy streets and the clean and pure well-to-do home. She was made to speak of moral degeneracy, associated with the slum and dirt and disease and so on. And she was, thereby, disassociated with the suburb and the clean white bourgeois man (some of whom paid for sex). Unsurprisingly, the prostitute became the subject of surveillance and punitive regulations designed to identify and classify her, remove her from the streets and improve her (see Walkowitz, 1980). Cities and bodies were, and are, part of a disciplining visual regime in which sexuality was, and is, the reason to look and, thence, to (not) touch: but there can be no easy looking or touching while exchanges in the contact zone between classes, genders and races involve punishment (whether by 'contamination' or regulation).

Stallybrass and White's work shows that particular social and psychic spaces were organised through Imaginaries which involved specific sites such as sewers, slums, pigs, bourgeoisie and prostitutes. These sites were assiduously mapped onto a simultaneously real, imaginary and symbolic city; where the city and the sites were rendered visible and invisible, significant and insignificant, through the internal relationship between power, desire and disgust. At the same time, power, desire and disgust were feverishly mapped onto bodies and spaces through the social distinction of categories of 'high' and 'low'. The internal relationship between 'high' and 'low' meant that these categories had to be continually named, watched and policed; it also meant that they were continually transgressed – socially and spatially. Finally, Stallybrass and White demonstrate that bodies and cities are geographies of meaning, identity and power which are mobile, conflictual and local.

In summary, topographies of mind, body and city, while not being reducible to one another, are mapped through citation of one another; just as topographies of subjectivity, meaning and power – such as class, gender, race, sexuality and so on – are mapped through resonance and dissonance with one another. Such an understanding exemplifies a 'psychoanalysis of space', of bodies and cities, because the techniques of this cartography are modalities of identification and desire (see Chapter 4 above).

BODIES AND CITIES: GEOGRAPHIES OF SUBJECTIVITY, MEANING AND POWER

In the next two chapters, I will explicate specific topographies of the body and the city in order to disclose the specific ways in which the social and the psychic are traced through space by the internal dynamics of power, desire and disgust. 'Mapping topographies' may be a nice geographical metaphor for a 'psychoanalysis of space' but there are three caveats to make at this stage. First, topographies drawn by different power relations in different places and at different times should not be mapped one-to-one with others: the sites I have chosen illustrate specific relationships at specific times in specific places, so 'general' conclusions have to be drawn out carefully. Second, the notion of a topography is not intended to ignore or deny historical shifts in relationships, but to highlight the playing out of historical relationships in specific spaces. And, finally, deploying a spatial metaphor like topography is not intended to suggest that somehow these 'geographies' exist prior to their constitution in maps of meaning, identity and power; there are no 'essences' to be mapped. Instead, this analysis traces shifting relationships between the individual and the social, where each is constituted as an effect of the other, but where neither is reducible to the other nor understandable solely in terms of the other; such that, for example, any individual may be placed in a particular place in relation to class, gender, race, sexuality, political orientation and so on, but aggregating these positions would not in all cases tell the analyst how a person would feel, think or behave in any given situation (see also Part I above).

Towards the end of Part II, I argued that it is necessary to extend the understanding of subjectivity and spatiality to incorporate the constellation of relationships within which subjects find a place in the world, to draw on the ambiguity of 'sanction' in order to convey a sense of the fixities and fluidities of subjectivity, to elaborate the partialities of subjectivity, to situate the truths of subjectivity simultaneously within repeated, stylised social practices and personal meaning systems, to differentiate between dissimilar, though interconnected, spatialities, and to disclose individual, group and social maps of power and resistance. These resources, it should be added, need to be selectively drawn on in any analysis; they also should not be seen to exhaust an analysis. In the chapters which follow, I will selectively deploy these ideas to contribute to what might be called a 'psychoanalysis of space' or an interpretation of the 'psychodynamics of place', depending.

This 'psychoanalysis of space' differs from Stallybrass and White's in one other key respect.[9] As I have shown, their analysis discloses the ways in which the internal dynamic of power, desire and disgust operates through internally-related bi-polar oppositions (such as bourgeoisie versus proletariat, classical versus grotesque bodies, male versus female, clean versus dirty, black versus white and so on). They insist that these oppositions can be 'transgressed' precisely because they are internally related. The notion of 'transgression' which they are working with is an inversion of categories of 'high' and 'low', such that the 'low' crosses the boundary between high and low and appears in the place of the 'high'. However, in parenthesis, they admit that 'there is another, more complex use of the term which arises in connection with extremist practices of modern art and philosophy; these designate not just the infraction of binary structures, but movement into an absolutely negative space *beyond the structure of significance itself*' (1986: 18).

In the readings which follow, I am interested not just in the ways in which power, desire and disgust operate to maintain bi-polar distinctions and the ways in which those categories are internally related, but also the ways in

9 As with Stallybrass and White, my perspective resonates with, and differs from, that offered by Elizabeth Grosz (1994).

which other 'negative spaces' are opened up, shut off or closed down in the mobile and conflictual fusion of these interrelationships. The term I propose to use for any 'negative' space which lies 'beyond the structure of significance' is 'third space' (see Pile, 1994b); third spaces do not simply lie beyond dualisms, they call into question the constitution of dualisms; third spaces are not simply gaps between axes of power (such as race, class, gender, sexuality), they are also created out of the interactions between different power relations, different desires and different fears; third spaces are also inflected in geographical space – in the body and in the city.

It is now possible to move on to Chapters 6 and 7, which will pick out specific 'contents' – to use Stallybrass and White's expression – constituted through the body and through the city (some of which have already been hinted at in this introduction): in particular, bodies are taken to be an intensifying grid of power and identity; likewise, cities are a spectacle of buildings, crowds and figures: the profusion of forms charms, fascinates and occupies observers of the city in specific ways – in these ways, it is possible to find a psychodynamics of place.

6

BODIES

desire and disgust in the flesh

> There is an immediate relationship between the body and its space, between the body's deployment in space and its occupation of space. Before *producing* effects in the material realm (tools and objects), before *producing itself* by drawing nourishment from that realm, and before *reproducing itself* by generating other bodies, each living body is space and has space: it produces itself in space and it also produces that space.
>
> (Lefebvre, 1974: 170)

INTRODUCTION

Henri Lefebvre notes, in *The Production of Space*, that the body both produces itself in and produces space, that each body is deployed in and occupies space and that each body is and has a space. The relationship, he implies, is immediate: there are no mediating terms between the body and the space it is, is in or has; the body does things, sustains itself and procreates. From this perspective, Lefebvre usefully undercuts any understanding of the body which abstracts it from its space. However, there remain problems with his formulation. Not only do such observations sit uneasily alongside Lefebvre's own treatment of space, but they also appear to be remarkably ahistorical, asocial and apolitical. Nevertheless, the body is an appropriate place to begin a 'psychoanalysis of space', partly because it is one site for the intensifying articulation of power, desire and disgust, of the individual, the social and the spatial. Moreover, the body is never merely a passive surface, a leaky container of visceral fluids, a collection of orifices, limbs, feelings, organs, and so on.

In this chapter, I will work through two different (though related) examples of body-space in order to unpack the specific kinds of mediations which belie the immediacy of the body and space: the first is an examination of Freud's notorious use of the phrase 'dark continents' to describe adult women's sexuality and the second is a discussion of Klaus Theweleit's interpretation of male fantasies. There is a problem with this strategy: each example permits the elaboration of specific aspects of the ways in which the space of the body might be psychoanalysed. It is a little too easy just to start with the health warning that these studies are meant to be illustrative, rather than exhaustive. So I would like to introduce this chapter, briefly, through the work of Mary Douglas (1966), who like Lefebvre also seems to think of the body as having a space and being a space. While there are obvious problems with the study's 'anthropo-pathologising' of the other and its lack of attendance to different modalities of power, what is striking is the sense that bodies are spaces where multiple, interrelated meanings can be mapped. This discussion is intended both to set up the two substantive studies and also to suggest their specificity – but is intended also as a reminder of some of the co-ordinates which I have set out for a psychoanalytic cartography of the subject. The proposition, here, is that the body is both the mind and the body, the personal and the social, and more: in Freud's term, it is 'body-ego' – a term which also implicates its temporalities and spatialities (see Brennan, 1992).

The body, it seems, is open to multiple writings and readings. One version of this is psychoanalysis, but this is not Mary Douglas's frame of reference. Indeed, she criticises Norman Brown's psychoanalytic interpretation of primitive cultures (pages 117–127; see Brown, 1959). She complains that he treats primitive cultures as if they were infants or retarded adults. Douglas refuses to treat these cultures as if they were entirely the product of infantile anxieties or to treat sublimation as if it were the mark of a people's civilisation or development (however 'civilisation' or 'sublimation' is valued). Instead, she analyses the ways in which the values of culture are inscribed on the body, and not just the surface of the body. One example she uses is that of the Hindus. According to her, the caste system in Hindu society marks out certain bodies as belonging to the head of society, while others are at the bottom; in this social division, those at the head do the thinking, have the control and perform the rituals, while those at the bottom are simply there to carry away the detritus of society. In this way, she suggests, the body reflects society and society reflects the body: each is mirrored in the other.

From this perspective, threats to the body politic are symbolised as threats to the body. Thus, she states that 'rather than oral or anal eroticism it is more convincing to argue that caste pollution represents only what it claims to be. It is a symbolic system, based on the image of the body, whose primary concern is the ordering of a social hierarchy' (page 125). For me, it is curious that she does not argue that threats to the body, some of which are experienced in childhood, are written into the symbolic system. Instead, the body appears as a basic scheme, or site of anxiety, which is given meaning by politics, society, culture. She suggests that

> body is a model which can stand for any bounded system. Its boundaries can represent any boundaries which are threatened or precarious. The body is a complex structure. The function of its different parts and their relation afford a source of symbols for other complex structures. We cannot possibly interpret rituals concerning excreta, breast milk, saliva and the rest unless we are prepared to see in the body a symbol of society, and to see the powers and dangers

credited to social structure reproduced in small on the human body.
(1966: 115)

For Douglas, the body is traced with the values of culture: the contours of the body are the contours of society: each reproduces the 'nature' of both the powers and dangers credited to social structure. Even so, fears for the flesh commonly cluster around sexuality (page 157). It is sexuality and, in prevailing economies of meaning and power, women's bodies which tend to bear the burden of these fears. Thus, it is female purity which is carefully policed, and women's sexual transgressions which are brutally punished (page 125). She also demonstrates that sex is given a special status in society and that institutions of power commonly rely on sexual differentiation, where men hold the power, regulate the law and control meaning (page 138 and Chapter 9). However, she boldly states that there can be no simple psychoanalytic interpretation of this preoccupation with sex, female purity and the woman's body. So, while rituals draw on personal (bodily) experience, they are not an attempt to cure psychoses or neuroses. Instead, 'rituals work upon the body politic through the symbolic medium of the physical body' (page 128).

The social and bodily rituals of blood, skin, excreta, parturition, and so on, mark out the spaces of a body-ego (and a not-body-ego) which, despite Douglas's objections, I would argue can also be understood psychoanalytically. Thus, the 'interplay' of form and formlessness, or of internal and external, might be seen as the direct result of particular, culturally-embedded, childhood experiences of the mirror stage (following Lacan). While agreeing with Douglas that neither the social, the political, the economic, and the cultural, nor the bodily and the personal can be reduced to infantile experiences, it is surely within Douglas's purview that the Real body must also write rituals through both the inarticulate margins of the Imaginary and also articulate power in the form of the Symbolic (see page 98).

Although there are many ways in which it might be possible to plot alternative maps of the (body politic) culture/nature (physical body) story using psychoanalysis,[1] I have outlined some keys areas which I am interested in:

1. the relationship between 'objects',
2. the repeated hiding of things in the open, though in disguised forms,
3. living in a hall of mirrors, in shadows, echoes and mirages,
4. the masquerading of self which defines the subject,
5. real, imaginary and symbolic spatialities, and
6. the placing of the subject within multiple relations of power and resistance.

---

1 For recent studies in anthropology which are informed by psychoanalysis, see Heald and Deluz, 1994.

It is possible mechanically to read off these concerns from Mary Douglas's work (although I have no intention of doing this in each of the case studies which follow). One reason for doing such a reading, here, is that it helps exemplify each of these themes. *First*, she notes that there are distinct sets (constellations) of object relations, from those nearest the body to those located within the social structure, where these fleshly 'scales' are simultaneously personal, social, physical and symbolic. Thus, for example, she argues that, in particular situations, individuals 'behave as if moving in patterned positions in relation to others, and as if choosing between possible patterns of relations' (page 100). *Second*, she shows that social rituals act through bodies not only to repeat unspoken anxieties but also to allay those fears by 'symbolising' them (see pages 170–171). *Third*, the body becomes a mirror of society, in which people can find where they are in the social structure (see above and, for example, page 121). *Fourth*, through following social scripts of the body, the 'subject' is both brought into existence and given a (moral) way of acting in the world (see Chapter 8). *Fifth*, throughout the book, she demonstrates that the flesh is a point of capture for multiple, real-and-imagined-and-symbolic spatialities (and temporalities), so for example initiation rituals create adult bodies which are located in adult spaces, while pollution beliefs periodically mark bodies as dirty until cleansed, and heads and hearts symbolise and spatialise different experiences of the body-ego (while not mentioning the phallus). Moreover, Douglas supposes that the body is the container for the internal and is separate(d), therefore, from the external by boundaries (compare Chapters 7 and 8). *Finally*, the example of the Hindus shows the ways in which power is plotted into the allied division of the bodily, the social and the spatial: the élite is the head which resides in specific sites, while the masses are differentiated parts of the body, which subsist in their proper places.

In this narrow rereading, it seems that Douglas has already provided a psycho-geography of the body. However, as I said, her work is not psychoanalytic: she does not share Freud's understanding of the 'unconscious' nor Lacan's sense of mirroring, nor are her 'rituals' analysed as 'repressions' or 'repetitions', nor is culture built by 'sublimation', nor does she have a psychoanalytic sense of object relations (broadly conceived to include everything from transference to identification). Finally, I would argue that the space of the body is more than the double of the personal and the social or of the internal and external: it is both and more – it can also be a 'third space'. Altogether, of course, these are not so much criticisms as suspicions, suspicions which I hope will not disappear in the course of this chapter.

To move on, although it is unfair to pick out two recurrent phrases from a book written thirty years ago and jump on them, the expressions 'primitive universe' and 'external boundaries' remind me not only of a particular way of thinking about 'civilisation' and bodies but also of the imagery of this thinking through geography. Instead of condemning Douglas's use of the term 'primitive' to denote other cultures or her application of the internal/external dichotomy, it is possible to consider an analogous geographical imagination.

DARK CONTINENTS AND GEOGRAPHIES OF THE BODY

At the beginning of Chapter 4, a connection was made between a geographer's, a writer's and a psychoanalyst's map of the mind. This map of the mind is not innocent of its corporeality because it is simultaneously a map of the body: that is, this cartography relies on the simultaneously social and spatial position(ing) of particular bodies in the map in(to) mapped and unmapped, mappable and unmappable places.[2] David Lowenthal was shown to selectively quote from Aldous Huxley's *Heaven and Hell*, thus:

> our mind still has its darkest Africas, its unmapped Borneos and Amazonian basins ... A man consists of ... an Old World of personal consciousness and, beyond a dividing sea, a series of New Worlds – the not too distant Virginias and Carolinas of the personal subconscious ... the Far West of the collective unconscious, with its flora of symbols, its tribes of aboriginal archetypes; and, across another, vaster ocean, at the antipodes of everyday consciousness, the world of Visionary Experience ... Some people never consciously discover their antipodes. Others make an occasional landing.
>
> (Huxley, 1956: 69–70; cited by Lowenthal, 1961: 249)

For Lowenthal, it seems, this imagery revealed the need to launch a geographical expedition to discover the as yet unexplored continents of the mind. Implicitly, if the darkest Africas, unmapped Borneos and Amazonian basins of the mind were to be opened up, then a specifically geographical imagination was required. I have already noted that Huxley's, and by extension Lowenthal's, use of this particular geographical metaphor to describe the mind resonates with Freud's phlegmatic observation that 'the sexual life of adult women is a "dark continent"' (Freud, 1926a: 124). In this subsection, I would like to take a closer look at the bodies implicit in this geographical imagination, certain aspects of which seem to be shared – somewhat surprisingly – by a geographer, a writer and a psychoanalyst. It is appropriate to begin the first substantive chapter of Part III with a closer look at this 'coincidence' and its possible ramifications for thinking about bodies, spaces and subjectivities for, as Haraway pointedly observes, 'bodies are maps of power and identity' (1990: 222). One starting point is Montrelay's inquiry into psychoanalytic understandings of femininity (Montrelay, 1970). She finds that the 'dark continent' metaphor tells of Freud's desire to find out about the unmapped femininity, and of his terror that it cannot be mapped (pages 260, 263 and 265) – such an analysis can be extended beyond the spaces of femininity.

The 'dark continents' imagery works to weave together specific accounts of race and civilisation into a tapestry which represents a specifically white

---

2 On mapping the subject, see Pile and Thrift, 1995b.

and western perspective. This perspective has it that it is possible to explore and civilise – by colonisation – Africa, Borneo and Amazonia; not just by making the occasional landing, but by staying and living there – as if they would be welcomed with open arms. In this exhortation, the brave explorer must continually appropriate new territories. It is hardly a leap of faith to suggest that this geographical imagination is located within a colonial discourse: that is, within a discourse which imagined, visualised, fantasised and rationalised its encounters and exchanges with 'others' in very particular ways.[3] According to Young (1995), colonialism did not just extend European power across the seas, it also meant that white bodies came into close, very close, contact with the bodies of others (see also Bhabha, 1986: 72). The 'mixing' of these bodies threw Victorian theorists of race into an ambivalent state of horror and fascination: they could not get their minds off the idea of people having sex, especially people of different 'races'. As Young shows, the multiplicity of terms for the 'hybrid' children of mixed race was an unavoidable testament to inter-racial sex – and desire.[4] The fundamental horror was that European race would degenerate if 'white' bodies were hybridised with the bodies of others, while the inner fascination was with the desirability of the bodies of those others.

Young argues that every time discussions about hybridity (and its synonyms) are entered into by white bourgeois Victorians, they betray the ways in which race and desire are inflected one in the other. For these Victorians, theories of race are also theories of history: race, and sex, are the motors of history – the rise and fall of civilisations were, for them, measured in drops of blood: in these theories, pure white blood was the source of civilisation, while the blood of others was (essentially) inferior and would 'adulterate' white blood if 'mixed', and 'miscegenation' would inevitably lead to the degeneration of civilisation. As Douglas argues, 'purity is the enemy of change, of ambiguity and compromise' (1966: 162). Miscegenation, from this perspective, required desire – it required people of different races to have sex with one another, irrespective of the circumstances within which that sex took place, one at least had to desire the other: thus, for example, even where the white man's rape of the black woman was explained by the inequalities of power in the master/slave relation, the eroticisation of that relationship

3 There is a substantial, and growing, literature on 'the colonial imagination'. The classic text is Edward Said's *Orientalism* (1978) – see also Said, 1993; Bhabha, 1994, especially 80–84; Spivak, 1988; and Young, 1995.

4 Unfortunately it would be too much of a digression to review Young's argument in full, but I would like to footnote the following point. While we might be unsurprised to learn that the white civilised man was seen as the source of all 'culture' (as indeed certain racial theories which asserted the supremacy of Caucasians and Aryans had it), it may be a little disconcerting to find out that the distinctiveness of Englishness was celebrated in the following terms (all cited by Young, 1995: 17): 'That Het'rogeneous Thing, An Englishman' (1703, Daniel Defoe) or 'a mongrel breed' and 'a hybrid race' (1861, *The London Review*). So much for the bulldog breed; apparently the English are a 'mongrel half-bred race' (1863, Carl Vogt). Race was simultaneously socially and spatially constructed, mobilised through simultaneously fluid and fixed hierarchies of binary oppositions.

still needed to be accounted for. The important point for Young is that whenever inter-racial sex emerges in these theories, it disrupts any possibility of the immutability of racial, sexual and cultural identity.

While these Victorian racial theorists were unable to produce accounts of race and desire which were fixed in the solid ground of pure, essentialised, autonomous identities because of the flow of desire across the borders of race, they nevertheless set up fixed, polarised, hierarchical and racist economies of difference – such that white men were placed at the highest point of (social, bodily) evolution, while others were placed on steps further down the ladder or even on different ladders altogether. In each theory, these economies of difference were given specific histories and geographies: for example, Johann Gottfried Herder established a seemingly egalitarian geography in which every culture had its place and its history, but in his racist scheme white civilisation was more advanced than the cultures elsewhere (see Young, 1995: 36–43).[5]

In these categorisations of race, the 'distance' between European civilised man and his racialised others was simultaneously bodily (blood, skin, brain, bones), social (civilisation, culture) and geographical (continents, nations, countries, tribes).[6] In this way, geographical categorisations were also hierarchical configurations of race, sex and civilisation – and it is in this context that Lowenthal, Huxley and Freud use their continental imag(in)ary. In his version of the metaphor, Freud deploys this hierarchical chain of associations of race, sex and civilisation to chart women's sexual life (in the singular!). Feminists have been angered by the 'dark continent' trope at least since Montrelay's inquiry (1970), but in this section I will rely on Doane's analysis (1991). First, it is necessary to look at the context within which Freud finds it appropriate to use the phrase, partly because it does not occur in one of Freud's major writings on femininity (as Doane points out, 1991: 209).

In his 'The question of lay-analysis' (1926a), Freud stages a dialogue between himself and an impartial person in order to describe the basic tenets, radical advances and remaining problems of a specifically psychoanalytic understanding of mental life. The battle between Freud and his imaginary protagonist is not easily, or even often, won. Nevertheless Freud never gives up the crusade to show the importance of a psychoanalytically trained 'band of helpers for combating the neuroses of civilization', finally attempting to

5 In this vein, Said points out that 'Geography was essentially the material underpinning for knowledge about the Orient. All the latent and unchanging characteristics of the Orient stood upon, were rooted in, its geography' (1978: 216). It should, therefore, be disconcerting to find out that Herder is seen as one of the three founding 'Fathers' of the modern academic discipline of Geography (see Livingstone, 1992, Chapter 4; see also Driver, 1992).

6 This observation also helps situate Freud's discussions of civilisation and the primal horde: Freud draws on these racialised understandings of the relationship between civilisation and culture to open out his theories of civilisation, repression and sublimation. Locating the racialised haunting of psychoanalysis has far-reaching consequences for the ways in which it is necessary to unpack and reconstruct psychoanalytic concepts.

disarm the reader's scepticism by humbly acknowledging that his smiling, fictional disputant does not believe in the value of this 'new kind of Salvation Army' (both page 170). During this ultimately vain struggle, Freud moves from the arena of his theories of neurotic illnesses (Section III) onto the terrain of sexual life (Section IV); it is into this section that he will drop the notorious 'dark continents' comment. Freud begins the section by asserting the value of psychoanalysis's unblinking and unyielding look into the sexual experiences, however nasty, of people on the grounds that 'sexual life is not simply something spicy; it is a serious scientific problem' (page 120). It is a 'scientific problem' (and therefore a psychoanalytic question), first, because decisive and formative moments in the patient's life are to be found in their childhood struggles with the power of sexuality and a serious problem, second, because 'our civilization is built up entirely at the expense of sexuality' (page 121).

Freud proceeds to mark his disappointment in the medical profession's denial of children's sexuality, but quickly observes that 'the sexual life of children is of course different from that of adults. The sexual function, from its beginnings to the definitive form in which it is so familiar to us, undergoes a complicated process of development' (page 121). There are three reasons for presenting this argument. First, to note that children have one sexual life, as do women in Freud's 'dark continents' throwaway line. Second, to observe that the sexual function in adults is 'so familiar' to Freud (and the impartial person), in direct contradiction to his avowed puzzlement over femininity (here and elsewhere). Possibly this is not so curious – Freud might only be referring to the development of boys and men (although there is no textual evidence for this, Freud does not differentiate between girls and boys here as he does later), or he might only be making a rhetorical point in the face of the impartial person's scepticism. Lastly, to mark Freud's developmental understanding of sexual life. Although it is possible (he says) to find out about anatomy and physiology in medical schools, any archaeology of sexuality is incomplete without 'a familiarity with the history of civilization and with mythology' (page 122). In this argument, it would appear that psychoanalysis's contribution to the understanding of mental life is to give equal weight to the development of culture (concerning civilisation's discontent with sexuality) and mythology (the main figures are Narcissus and Oedipus) as to the cold-blooded anatomical facts.

Strange mythological and fairy tales, Freud argues, become intelligible as a by-product of psychoanalysis. Thus, stories about ravenous wolves represent children's fear of their fathers (see Angela Carter on this, 1979). In particular, Freud finds that 'male children suffer from a fear of being robbed of their sexual organ by their father, so that this fear of being castrated has a most powerful influence on the development of their character and in deciding the direction to be followed by their sexuality' (1926a: 123). Once more, he turns to mythology in order to substantiate this claim. Freud (twice) relates the story of Kronos (Time) who not only swallowed his children but also his father Uranus. Later, Kronos is punished by his son, Zeus, whose mother saved him from Kronos's cannibalistic infanticide. Such myths, Freud argues, re-present the constellation of object relationships in early childhood sexuality.[7] Thus, an implicit connection is made between archaic myths, archaic sexuality and archaic (Hellenic) civilisations. Freud soon renders the metonymic chain of associations explicit – his diagnosis is this:

> in the mental life of children today we can still detect the same archaic factors which were once dominant generally in the primeval days of human civilization. In his mental development the child would be repeating the history of his race in an abbreviated form, just as embryology long recognized was the case with somatic development.
>
> (Freud, 1926a: 124)

I will make three observations, quickly, before moving on through his narrative. First, Freud slides from associating the mental life of all children with archaic forms of civilisation to a position where (only) boys act out the history of their race. Second, there is an explicit reference to race, although the exact relationship between Greek mythology and the history of other European races is not spelt out (see Young, 1995, on this point). Third, and following from the first two points, in drawing analogies between childhood development and mythological settings, Freud abstracts both the myths and children from their historical and geographical circumstances. The scene is now set for 'dark continents'' walk-on part.

Having dealt with the archaic forms of both childhood sexuality and civilisation, Freud proceeds to talk about the place of the 'proper' female sexual organ in the development of early infantile sexuality. Freud has no doubt: it plays no part in it because it cannot be discovered (being on the inside), whereas the male organ (being on the outside) is there for all to see – apparently. All children's interest is, he argues, directed to the absence or presence of the penis, which thereby becomes the one signifier of sexual difference (see Part II above). That is, children's concerns cannot circulate around the vagina, he suggests, because it is not there. Uncannily, Freud's inability to find the vagina in early infantile sexual life 'returns' as an admission that

> we know less about the sexual life of little girls than of boys. But we need not feel ashamed of this distinction; after all, the sexual life of adult women is a 'dark continent' for psychology.[8] But we have learnt that girls feel deeply their lack of a sexual organ that is equal in value to the

---

7 It should also be noted that the Victorians also drew on other geographical imaginations to write fairy stories which spatialised the child through a sense of interiority, see Steedman, 1995.

8 Interestingly, the expression 'dark continent' was in English in Freud's original German text. It does not stretch the imagination too much to suggest that it is in English because of its association with the British Empire. In particular, Africa was seen as the 'dark continent' (as seen in the extract from Aldous Huxley); dark only because of the significance given to the colour of the skin of the peoples found there by white explorers – often funded, sponsored and aided it has to be said by the Royal Geographical Society.

> male one; they regard themselves on that account as inferior, and this 'envy for the penis' is the origin of a whole number of characteristic feminine reactions.
>
> (1926a: 124)

These characteristic feminine reactions are not spelt out. Indeed, 'femininity' is not the object of this piece of rhetoric, as articulated through Freud's struggle with his imaginary adversary. Instead, the point he is making concerns the importance of castration for both boys and girls and the deep feeling of penis-envy in girls alone. From here, he talks about the need in children to appease anxieties associated with their early sexual life. This discussion leads into a description of the 'Oedipus complex' in boys, the incest prohibition and Julius Caesar's encounter with Cleopatra in Egypt (pages 125–126).[9] These stories are deployed in order to provide legitimacy for Freud's theories simultaneously by analogy, by revelation and most importantly by tapping into the social values attached to classical learning: that is, in Stallybrass and White's terms (1986), by transcoding psychoanalysis with high culture. This trafficking of values is important – it implies the circulation of other, especially racial and sexual, meanings.

Freud would appear to be 'in denial' (as pop psychology calls it): he argues that 'we need not feel ashamed' that 'we know less about the sexual life of little girls than of boys', because 'after all' he does not know that much about 'the sexual life of adult women' either. And he covers his shame in the cloak of generalised ignorance about the 'dark continent'. It would appear that both Freud's feelings of shame and his choice of cloak stem from the same root, that a white bourgeois Judaeo-Christian imperial adult male imagination cannot conceive of its others – in their racialised, sexualised, gendered, infantilised, and so on, bodies. The point here is that these bodies lie unseen in the contact zone between this imagination and its others. Thus, what interests Doane is the quality of the associations which Freud makes between 'dark continents' and female sexuality (1991).

In her reading, Doane draws out the specific ways in which this master-narrative is embedded in interanimated relationships of race, sex and gender – and in this analysis she discovers that Freud's geographical metaphor describes a specific field of otherness. She argues that 'the phrase transforms female sexuality into an unexplored territory, an enigmatic unknowable place concealed from the theoretical gaze and hence the epistemological power of the psychoanalyst' (1991: 209), and that 'a metonymic chain is constructed which links infantile sexuality, female sexuality and racial otherness' (page 210), where for Freud 'the "primitive" is remote in time, it is the "childhood" of modern man' (page 211). The 'immature', feminised, racialised bodies of others – children (especially girls), women, 'primitives' – are distant in time,

---

9 See Young, 1995, Chapter 5, on the 'whitening' of Egyptian civilisation in racialised western knowledges.

if not in space, but actually also in space. These are not the only 'others' which Freud uses in this way. Thus, in Gallop's analysis of Freud's interpretation of sexual jokes, she shows that Freud links the 'lower classes' to women by associating them both with the infantile (Gallop, 1988: 38). Thus, an ambivalent, intricate and dynamic economy of meaning produces, circulates and exchanges the values of power. In this economy, embodied others are simultaneously associated, denigrated *and distanced*.

Despite this distancing, these others nevertheless act as a repressed trace within the site of exclusion. In this case, 'modern man' is a grown up 'primitive': the primitive is simultaneously past and present – like anybody's childhood. There are three notable effects of this distancing. First, the distance permits the theoretical gaze of the psychoanalyst to focus on these others. Second, this distance prevents the psychoanalyst from seeing the differences amongst and between those others which he (Freud) notices and, indeed, enables him to be indifferent to their difference. Third, the material or truth effect of this epistemological distance is that these others can be seen as 'other' while, as other, they become fascinating. Thus, the site/sight of these (infantilised, feminised, racialised) bodies become points of capture for (white, male, adult, highbrow, Judaeo-Christian, western) fantasies of horror-and-pleasure.

Doane, working within the frame of film studies, is interested in questions of visibility and representation: in this, the body – or more precisely parts of bodies – is always implicated. In her analysis of a number of films, especially *King Kong* (1933) and *Birth of a Nation* (1915), she shows how the trafficking metaphor of 'dark continents' plays into an intersection of the erotic and the exotic in white western cinema.[10] In the representational practices and discourses of cinema, specific bodies came to connote excess, and especially excessive, sexuality – the native, the savage, the woman and the monster.[11] For Doane, a convoluted economy of subjectivity, meaning and power catches these bodies: for example, the white woman was often seen to be unknowable and sexually excessive and thereby associated with blackness, while elsewhere white women became the repository of racial purity and their civilisation was contrasted to the excesses of 'primitives' and, especially, 'native' women (similarly see Bhabha, 1986: 74).

In this contorted gymnastics of signification which constructs women as other and lower to men, and blacks as other and lower to whites, and children as other and lower to adults, and the primitive as other and lower to the civilised, it is easy to find support for Stallybrass and White's assertion that the body is an intensifying grid for the power geometrics of transcoded binary oppositions. The difficulty for such geometrics is that in some places certain bodies have to

---

10 On the intersection of 'race' and 'sex' in these films, see also Snead, 1994, Chapters 1, 2 and 3.

11 Work on the 'hypersexualisation' of particular (raced, sexed) bodies, and the ambivalences of desire and disgust that this constitutes and is constituted by, can be found in many places (see for example Mort, 1987: 186; Mercer, 1994: Chapter 5; Bruno, 1993 and Creed, 1993).

be higher (such as the 'pure' white women) and in other places lower (such as the 'fallen' white women) depending on the situations within which the body is being produced, reproduced and exchanged.

The topographies of these bodies extend far beyond the abilities of the codified maps of difference to contain them. Doane uses two examples to illustrate this 'excess' spatiality: first, Fanon's psychoanalysis of 'the grotesque psychodrama of everyday life in colonial societies' (in Homi Bhabha's memorable words, 1986: 71);[12] and, second, the hegemonic visual regime of cinema. She exemplifies her argument through an analysis of two films: D. W. Griffith's *Birth of a Nation* and Douglas Sirk's *Imitation of Life* (1959). I would like to elaborate her psychoanalytic understanding of the intersection at body-spaces of discourses of race, sex and gender by thinking through her discussion of these films. In this film, the body-space becomes marked by specific characteristics which are taken to lie at the heart of identity: skin, blood, sex.[13] However, this 'identity' continually slips away, partly across the constellations of relationships in which the figures are placed, partly through the masquerades of subjects (identity-as-masquerade, cinematic narrative as masquerade, audience as masquerade, and so on), and partly through the setting of the film.

The context for D. W. Griffith's *Birth of a Nation* is the North American Civil War (1861–5) and its aftermath. A central thread of the plot is the abolition of black slavery and it articulates and expresses its racist attitudes and anxieties partly by dramatically 'linking rape, lynching and castration' (page 227) and partly by celebrating the birth of the Ku Klux Klan. While the midwives (to follow Griffith's natal analogy) are the homologous struggles between South and North and between white and black, the child is 'a nation'. The arrival of this child is, in the racist rhetoric of this film, the delivery of the Ku Klux Klan: in this way, 'the nation' is identified with pure white men. The film attempts to stabilise and reproduce the supposed purity of whiteness through its narrative of encounters between the white Southern Cameron family and various (impure) others. In each encounter, what is at stake is the purity of family, race and nation and the stake itself is the white women. In the spaces of the body, family and nation are the traces not only of power, desire and disgust, but also of race, sexuality and class (see also Lebeau, 1995). For Doane, the body images in *Birth of a Nation* act like mirrors which establish identity (following her discussion of Lacan's mirror stage, page 225). She argues that in this film, white identity is grounded both

12 Although she agrees with Fanon (1952) that skin takes on an inescapable meaning in the visual regime of colonialism, Doane criticises him for overlooking the experiences and positions of black women within the psychodrama of colonisation and, instead, foregrounding white female sexuality (Doane, 1991: 216–222). Following Fanon, Doane shows that in cinema too whiteness regulates blackness by making the black skin signify corporeality (page 224). See also Chapter 8 below.

13 There are distinct resonances, here, with Bruce Pratt's (1984) poignant discussion of identity. Gillian Rose uses this article to exemplify paradoxical spaces of femininity (1993: 156–158). See also Chapter 8 below.

in its absolute opposition to and in its unadulterated difference from black identity, where blackness is located in an excessive, over-visible, over-sexed body: skin is identity, but only for the blacks; white identity is established by (invisible) blood. Even so, no simple reading of these reflections can be offered for, in the film, whites 'drag up' in white robes as part of Ku Klux Klan rituals and white actors 'black up' to play the parts of the blacks: that is, white bodies masquerade as white ('good') and as black ('bad') bodies.

Douglas Sirk's 1952 remake of *Imitation of Life* also has a 'confusion' of bodies and spaces. Hollywood movies of the period tended to represent black women only as servants (i.e. through racialised, sexed and classed relations), but the plot of the film builds a different (white) representation of black femininity. For Doane, Sirk's film enables an examination of the (fantasised, real) relationships between white and black women, but this constellation is played out through 'visibility, knowledge, power, and masquerade' (page 233). And the film is 'geographical': its racialised dynamic concerns the racial transgression of a woman who moves from one place to another – from the black inner city to the white suburbs.

The central character in the plot is a mulatta who 'passes' as white, but she is caught in an impossible space between two cultures, two places and two bodies – white and black. Thus, Doane argues, the woman's whiteness 'allows her economic and cultural opportunities which would otherwise be denied her, but grasping these opportunities means denying her own cultural heritage and familial connections' (pages 233–234). Nevertheless, the 'impossible' body of the mulatta is caught in a non-space between two distant race-identified 'continents'. To say that this distance between white and black is paradoxical is a simultaneously psychoanalytic, geographical and social point: it involves the oscillation across ambivalences of not only desire and disgust, but also of remoteness and proximity, and also renderings of race, gender, sexuality and class.[14] As she did for *Birth of a Nation*, Doane shows that the film is haunted by excessive, racialised fears such that the 'mulatta' is played – in a grotesque masquerade of visible racial signifiers – by a white woman (who is pretending to be a black, who is pretending to be a white!). Thus, in *Imitation of Life*, Doane shows that 'there is one body too much' because the film cannot let itself either imagine or represent a black woman who actually imitates white life.

In both films, there is a similar moment:

> When a white patriarchal culture requires a symbol of racial purity to organize and control its relations with blacks (usually black men but sometimes black women as well), the white woman represents whiteness itself, as racial identity and as *the* stake of a semiotics of power ... She becomes the norm rather than a limited racially defined being.
>
> (1991: 244)

---

14 See also Bhabha, 1983, 1984.

In this economy of desire and fear the resonances with Victorian theories of race and hybridity are unsurprising (e.g. it is not a surprise to find that a mulatta should be classed, by blood, as black – and not white) – if only because racial science took on a virulent and racist form in North America's deep south (see Young, 1995, especially Chapters 5 and 6). I would quietly add that both these films also demonstrate the simultaneity of psychoanalytic and spatial dynamics: the bodies mirror and are mirrored, are repressed and expressed, are placed in constellations of relationships (which make a difference), are masquerades, are constituted by and constitutive of power – and involve simultaneously real, imaginary and symbolic spatialities: the body, the nation, white domesticity, and so on. This is not, however, the end of the story; Doane adds another co-ordinate to this particular map of the subject.

With all this play-acting going on (as with any film, drama, etc.), Doane suggests that it is possible for the audience to take up different positions in relation to the characters, actors and bodies in the film. While these positions can be complex, ambiguous and ambivalent, the filmic mirror offers specific re-presentations with which the viewer can identify – or not. Viewers can make their own mind up about how they feel about any movie but both *Birth of a Nation* and *Imitation of Life* offer a plot, characters and actors who play those characters, which are circumscribed within a particular arena of subjectivity, meaning and power – which is mobilised through hierarchical and opposing values, themselves intoned in hierarchised and opposed bodies, spaces and subjectivities. For example, while white and black bodies, white and black spaces and white and black identities are fluid and displaced in each film, their positions are located, fixed and enduringly maintained in contradistinction to one another. It should be noted that each film also shows precisely how and where these dichotomised gradations of meaning, identity and power are disrupted, interrupted and ruptured: that is, when bodies, spaces and subjectivities cross the borders between the one and the other and thereby create unmapped and unmappable 'third' spaces.

Doane goes on to argue that the cinematic spectacle does not take place in abstraction (on spectatorship, see also Lebeau, 1995). It should be remembered that cinemas are social spaces, and that specific things happen in these places: 'The space of the theater becomes a space of identificatory anxiety, a space where the gaze is disengaged from its "proper" object, the screen, and redirected, effecting a confusion of the concept of spectacle' (1991: 227).

The screen is not a simple reflection, nor are the characters of any story to be read only one way, but nor can the production and reception of any film be abstracted from its setting – in part, as 'a space of identificatory anxiety'. Doane infers that 'the narrative of *Imitation of Life* ... thus seems to be very much concerned with producing and allaying anxieties about identity, position, social categorization' (page 238). In this respect, the fear that there is nothing essential about racial, sexual, gendered and classed identity provokes an identificatory anxiety which the cinematic experience 'fixes' by relocating the viewer's gaze within its real-and-imagined-and-symbolic grids of meaning, identity and power. Perhaps this is why women caught in the wrong place always pay the ultimate price – with their lives (Lebeau, 1995; see also below). From this perspective, films materialise truths about characters, their positions and their social context, which are embodied in necessarily specific ways. Thus, in the middle-class setting of *Imitation of Life*, the black servants serve as an embodiment of what white women are not or are no longer (page 240). The simultaneously social and spatial hierarchy of whiteness over blackness is affirmed through this proximate juxtaposition in the narrative space of the film. Indeed, spatial understandings of these relationships enables an awareness of the ways in which tones of bodies and subjectivities resonate and dissonate with each other. Thus, for Doane,

> the trope of the dark continent is an early symptom of the white woman's fundamental and problematic role in the articulation of race and sexuality. The nature of the white woman's racial identity as it is socially constructed is simultaneously material, economic, and ... subject to an intense work at the level of representation, mobilizing all the psychic reverberations attached to (white) female sexuality in order to safeguard a racial hierarchy.
>
> (1991: 245)

The production of the space of femininity, as exemplified by Freud's 'dark continents' metaphor, is unintelligible unless the reverberations of maturity, morality, intellect, civilisation, gender, sex, race, class and so on are traced out. And one site where these psychic resonances are captured and intensified is the white woman's body. This section has drawn on Young's analysis of colonial desire and Doane's film theory to trace out the psychoanalytic and spatial connections drawn and covered by the 'dark continents' trope. Nevertheless, the white imaginary has constructed these 'dark continents' of the raced, sexed body in other ways: related, but distinct, psychoanalytic and spatial inflections through the body are revealed by Klaus Theweleit in his extraordinary analysis of the fantasies of proto-Nazi men in inter-war Germany (1977, 1978). In particular, it is possible to say something more about violence and the body politic, while not losing sight of (in)visibility, ambivalence and spatiality.

MALE FANTASIES: KLAUS THEWELEIT AND THE BODY POLITIC

In the preceding analysis of his use of the phrase 'dark continents', I have boldly situated Freud within an imperial geographical imagination. I have put him on the side of white male bourgeois Judaeo-Christian culture, but Freud and his family were persecuted by the Nazis, who did not see them as being 'white' enough. Freud cannot simply be located on the side of culture, power and authority. Then again, it is not adequate simply to label Nazis as insane, barbaric or evil, because they have their own stories: Hitler, for example, was not always the Führer (for a psychoanalytic biography, see Fromm, 1973, Chapter 13). It cannot be forgotten that the Nazis viciously distanced themselves from their 'others' through their use of concentration camps but, if Adorno's argument that 'genocide is the absolute integration' and that 'Auschwitz confirmed the philosopheme of pure identity as death' is correct (1966: 362), it also becomes necessary to 'psychoanalyse the white terror': if only because the white terror was grounded in particular psychic forms and in specific alignments of the body, geographical space and the social order. From this perspective, analytical attention must be paid to the simultaneously real, imagined and symbolic spatialities of nascent barbarism, to think through the fearful distances created through the desires and disgusts of the body.

For Klaus Theweleit, the central problem concerns the relationship between fascist discourse, the bodies of men and their relationship to their world (Volume 1: 24). Theweleit argues that fascist discourse did not simply express a relationship between men and their world; it was double-sided: on the one hand, fascist discourse actively constituted men's relationships to the outside world; on the other hand, it developed as a re-presentation of men's relationship to the world, growing out of their experiences of their own bodies and their relationship to the bodies of others, located simultaneously in an internalised world and an externalised world. Fascist discourse speaks about bodies and their relationship to others, bodies also speak of themselves and their relationship to others through fascist discourse. Fascist discourse, fascist fantasies, fascist subjectivities, fascist bodies and fascist spaces are constituted and fixed through boundaries which mark inside and outside, internal and external, true and false, good and bad. These borders are, however, neither permanent nor impenetrable and, so, their maintenance requires a permanent state of discipline, vigilance and war; hence, fascism's fixation on territories of all kinds (see Volume 1: 418). Significantly, in his analysis of narratives of adventure, Dawson (1994) develops a similar understanding of the relationship between soldiers, heroic masculinities and nineteenth-century British imperialism.[15]

The material for Klaus Theweleit's analysis is drawn from the records left by key officers of the *Freikorps*. The *Freikorps* was a notorious body of irregular soldiers, a kind of underground army-in-waiting in inter-war

---

15 Dawson draws on Kleinian psychoanalysis to explore the ways in which masculinities were imagined and fantasised. Accounts of heroic men provided an internalised masculine landscape through which men experienced themselves and on which they acted. For Dawson (1994), there is an internal relationship between hegemonic masculinities, soldier's heroism, heroic adventures and the British Empire. Although Dawson is apparently unaware of Theweleit's work, there are striking resemblances in understanding, if not (psycho)analysis: Theweleit rarely uses Klein, while Dawson hardly uses Freud, let alone Deleuze and Guattari. Sadly for my purposes, while Dawson, like Said (1978, 1993), is sensitive to the geographical imaginations of imperial discourses and deploys a richly spatial language to understand the relationship between the psyche, masculinity and the external world, he has (unlike Theweleit) virtually nothing to say about bodies.

Germany, who were involved in fighting on the side of a number of fascist causes. The majority of officers were drawn from the urban and rural middle classes and many were to become prominent Nazis. Moreover, Nazi organisations such as the SA and the SS were to draw on the resources, both people and arms, built up by the *Freikorps*. The *Freikorps* were involved in suppressing strikes, murdering workers and assassinating leaders of left-wing parties. These armed male brotherhoods created their own forms of justice and administered the arbitrary results of their 'tribunals': the execution of their right-wing nationalism became known as the 'white terror'.

There are three basic phases in the history of the white terror, according to Theweleit: the first from the end of the First World War until an abortive nationalist coup in November 1923, when the *Freikorps* struggled and failed to control the Weimar Republic; second, there was a period of decline for nationalism between 1924 and 1928 while the Weimar Republic was relatively stable; and, third, there was a resurgence of right- and left-wing struggle from 1929 to 1933 as the Weimar Republic's economy and polity became weaker and weaker. For Theweleit, however, these periodisations are less important than the ways in which 'white terrorists' positioned themselves in relationship to others, particularly through the language they used to articulate those relationships. For this reason, he 'sets out to chart the textual landscape of fascist fantasies, the "territories" of a fascist male desire' (Turner and Carter, 1986: 200). He argues that there is a fundamental ambivalence in the ways *Freikorps* soldiers talked about themselves, their hopes, their fears, other people and the social order. Their writings, in their memoirs, novels, eyewitness accounts and journals, oscillated between intense fascination and cold indifference, between extreme violence and passive obedience, between fanatical desire and furious hatred. These ambivalences, for Theweleit, are both expressive of and constitutive of these soldiers' psychic forms, their bodies, their geographical space and their desire for a new social order.

Theweleit builds a complex model of the fascist psyche by *critically* drawing on Freud's (1923) id/ego/super-ego topology of the mind, on Deleuze and Guattari's (1972) account of desire and Mahler's (1972) understanding of the pre-Oedipal circumstances of psychotic children.[16,17] He refuses to pathologise fascist males, in particular by rejecting the coupling

---

16 Theweleit resists

> the prevailing tendency to bring certain determinate phenomena under the umbrella of a psychoanalytic concept – a procedure in which individual concepts are isolated from their contexts and evolutions within Freudian practice, from their relationships to very specific symptoms, from their functions within Freudian theory as a whole, and so on – this tendency leads to a number of somewhat arbitrary abstractions.
>
> (Volume 1: 56)

This argument parallels Fromm's dismissal of 'psychoanalytic ingenuity' (1973).

17 In contrast to Theweleit, Dawson's (1994) analysis of soldiers' masculinities draws heavily on Kleinian accounts of ego-ideals, fantasy and psychic splitting (see also Chapter 4 above).

of totalitarianism and homosexuality (Volume 1: 54–57; Volume 2: 306–346). Theweleit discovers that 'before these men became soldiers, they lived within impenetrable circles of refusals and prohibitions. A world of terror perpetrated on their own bodies, a world that demanded the suppression of the slightest resistance bubbling up within them. The punishment for any refusal to acquiesce was death' (Volume 2: 402). For Theweleit, *Freikorps* soldiers were made in part through their familial relationships, in part through their military experiences, including both the military academy and warfare, and in part through their social, historical and geographical circumstances. It is in the making of these soldiers that Theweleit finds their predisposition and susceptibility to fascist politics.

Following Mahler's analysis of individuation (Theweleit, Volume 2: 206–263), he argues that, as boys, the soldier males individuated themselves from their mothers, but without resolving either their desire for unity with the mother (for re-engulfment) or their fear of being swallowed up by the mother (of re-engulfment). The result was that these boys did not encounter the Oedipal situation in the way Freud described (see Volume 1: 204–228). Instead, the 'normal' situation for boys in inter-war Germany was that they developed 'not-fully-born' egos which both paid little heed to the father and also longed for incorporation into ever larger bodies – bodies of men. Thus, Theweleit argues that the prohibitions placed on boys in the family created a situation where their bodies were in a state of perpetual helplessness, pain and punishment.

Where, for Freud, the body-ego is derived from bodily sensations which are projected onto the surface of the body and integrated into bodily life, in these deathly circumstances the body-ego is formed by the displacement of uncathected libido onto the peripheries of the body, such that feelings cannot be – must not be – incorporated into bodily life. Where, for Lacan, the child laughably gets trapped in the illusion of mirror stage, these boys are caught by both their fear of and desire for fragmentation, dissolution and dismemberment. For Theweleit, the combination of defence mechanisms protecting these men from their fears of dissolution, which displaced feelings onto bodily peripheries, projected their fears onto the bodies of others and terrorised any expression of femininity, produced an ego incapable of *negotiating* the relationship between the body, the body of others and the external world (Volume 1: 204).

Through displacement, these soldier men's desires, and fantasies, were projected directly into their bodily practices and their relationships with others. According to Theweleit, soldier men 'transmuted' reality through their fantasies (Volume 1: 215–217): they cast their eyes over reality, invaded it, occupied it, appropriated it, transformed it and condemned it to death. The real source of white terror was the realisation of these soldier men's fantasies, the making real of their fantasies. And this white terror was grounded in the ambivalences of their desire. These desires played out spatially: their ambivalence was maintained by an oscillation between positions, by the segregation and distancing of feelings, by a withdrawal from the world and by a wish for emptiness, for empty spaces. In this way, these men never experienced, confronted or resolved their ambivalences: remaining both desiring and fearing engulfment, where this ambivalence could only be integrated in death, in the blood of others:[18] 'The primary desire motivating such men to commit themselves politically to fascism is their compulsive urge to create and preserve their own selves by engaging in the dynamic process of killing' (Volume 2: 380).

Although there is no straight line from the particular constellation of familial relationships to the death camps, these cross-hatched body-ego-spaces were too fragmentary, too displaced, and too vulnerable to stand up to the vicissitudes of their desires and fears and to the imperatives of larger, higher, authorities (see Volume 2: 266). By absolving the individual of responsibility and by defending the individual against internal and external threats, fascist discourse enabled soldier males to commit themselves to fascist politics.

Theweleit argues that what the soldier and the fascist want is war, where bodies of men are mobilised, where they can realise their potential as war machines. The ways in which these men defended themselves against their feelings played into these desires: they had learnt to protect their egos with a hard outer shell and now these men wanted to become 'men of steel', men of war. It is not difficult to interpret this desire as an (impossible) attempt by soldier males to make of themselves the phallus, as the indissoluble signifier (see Volume 2: 373). From soldier to fascist, these men safeguard themselves from ego-engulfment and ego-fragmentation by armouring their egos, by standing up, by standing firm, by becoming part of a larger body and by becoming parts of the war machine. These men's bodily boundaries extended to the corps, to the body of men, to them against the world (see Volume 2: 164). Fixing and maintaining the armour of the body and the integrity of the corps became an ineradicable component of the fascist soldier's identity, their desires and their politics.

> What fascism, therefore, produces is not the microcosmic multiplicity of a desire that longs to expand and multiply across the body of the world, but a *desire* absorbed into the totality machine, and into ego-armour, a desire which wishes to *incorporate* the earth into itself. This is the basis on which the typically fascist relation between desire and politics arises: politics is made subject to *direct* libidinal investment, with no detours, no imprints of mama–papa, no encodings through conventions, institutions, or the historical situation. Under fascism, the most common form of the 'I' is as a component within a larger totality-ego – the 'I' as 'we', pitted in opposition to the rest of the world.
>
> (Theweleit, 1978: Volume 2: 243)

18 This view of 'integration' and 'death' contrasts with that disclosed by Dawson (1994): although British adventurers lived by heroically risking death and necessarily died heroically, their masculinities were not forged in the desire for death, in the blood of others.

The armoured space of the fascist male body becomes a 1:1 map of authority, terror and death, absorbed into a total reality machine, where a sense of 'I' is lost to an 'us versus them' matrix of identity, where feelings are directly invested in the bodies of territorial institutions – such as the 'nation' – and in an embodied geographical world, involving sites like 'the earth', 'floods', and 'the city'. Theweleit shows that particular symbols, such as the nation, the earth, floods, the city, and so on, convey special intensity of feelings in these men's language. Whatever else is at stake in these feelings, Theweleit argues that they express an ever-present threat to these men's masculinity and manhood. For me, they therefore re-present the intersection of psychic forms, the body, imagined geographical space and the social (patriarchal, military, bourgeois) order. Following Deleuze and Guattari (1972), Theweleit argues that the important point is that the production of desire and the production of the social are inseparable: that is, flows of desire are mapped simultaneously into sites of intensity of meaning, through the body and across the surface of the earth. From this perspective, the contents of fascist fantasies are not only constituted within historical and geographical circumstances but also constituted by the desiring-production of male fantasies (Volume 1: 434). 'The nation', 'the earth' and so on, then, are important not simply as ideas or as metaphors for something else, but as specific *territorialisations* of desire, the body, geographic space, and the social order.

In fascist discourse, 'the nation' had 'in the first instance nothing to do with national borders, forms of government, or so-called nationality' (Volume 2: 79; see pages 73–107). Instead, the nation was constructed as brotherhood of soldiers, in contradistinction to the Weimar Republic and the masses. The nation represented the highest elements of German culture and the German race, so that the 'nation' became a term which stood at the intersection of the social order (German culture), the (Aryan German) body and geographic space (Germany). It therefore became a key co-ordinate in fascist struggles to unify the German people, signifying the superiority and purity of Germans, German culture and Germany. In this rhetoric, the process of integration of the German people into a nation involved three related features: first, the suppression of differences within the German people; second, the creation of unity through obedience to, and dominance by, one leader; and, third, the securing – and later 'purification' – of geographic space: 'Ein Volk, Ein Reich, Ein Führer'. These fascists desired the rebirth of 'the nation' and the destruction of the masses who stood in the way. In their nationalism, 'the nation' is an inner spiritual unity waiting to be written on the face of Germany as much as it is an internalisation of Germany as an inner spiritual force. As one writer, Ernst Jünger, put it: 'what is being born is the essence of nationalism, a new relation to the elemental, to Mother Earth, whose soil has been blasted away in the rekindled fires of material battles and fertilised by streams of blood. Men are harkening to the secret primordial language of the people' (1929; cited by Theweleit, Volume 2: 88).

In Jünger's writings, his fascist sensibilities and the sensibilities of the nation are identified: their rebirth will be nourished by streams of blood on Mother Earth. Attacks on Germany are experienced by such men as direct attacks on their own bodies and Germany's defeat in the First World War is felt by these men as emasculation, fragmentation, dissolution and despair. In this way, the geographies of pain and desire are territorialised across the heart and mind of the fascist soldier: 'the army, high culture, race, nation, Germany – all of these appear to function as a second, tightly armoured body enveloping his own body armour' (Volume 2: 84); moreover, he 'loves the power that derives from his feeling of unity, a feeling of unity with "Germany" and with himself' (Volume 2: 86). And what holds the fascist soldier's mind, body and the nation together is the brutal suppression of the bodies of others, in the merciless struggle to achieve one people, one empire, one leader.[19]

Moreover, from Jünger's sentiments, it is possible to pick out further elements: the identification of 'soil' and 'Mother Earth' with the nation's elemental spirit, of 'a primordial language' with the people, and of 'blood' with men's sacrifice. So, as the forest grows from the body of the earth, the German nation was to arise from its sleep (Volume 2: 18 and 96). In this way, fascist discourse naturalises the nation and the German people by associating them with an elemental spirituality. However, soldiers were more ambivalent about the earth: on the one hand, burying themselves deep in the earth, they protect themselves from being ripped apart by bullets and shells (Volume 1: 239); on the other hand, 'the anthropomorphised body of Mother Earth is presented as the cauldron that is threatening the soldier's body with scorching floods' (Volume 1: 238). The earth simultaneously threatens to engulf soldiers and to protect them from engulfment. It is not surprising, given these men's double/d economies of desire and fear over the mother's body, that the earth should be anthropomorphised as Mother (see Volume 1: 205).[20]

Theweleit argues that war brings these men closer to the earth and to their own bodies – and that as the earth explodes around them and as they bury themselves within its hollows, folds and holes these men feel a thrill (desire and fear) and a power which cannot be matched. As the grenade explodes, ego boundaries and bodily boundaries dissolve, the explosion destroys the relationship between internal and external, between subject and object. The earth bursts within the soldier: the body is the earth. The interior is never only interior, and it is for this reason that the soldier's body can be imperilled by engulfment, by the flood of the masses, of the reds, of blood.

Men of the *Freikorps* were, according to Theweleit, without exception

---

19 This analysis supports Freud's interpretation of the strength of group bonds and the role of the leader, see Chapter 4 above. The important point is that Theweleit demonstrates just how these relationships are spatially and geographically constituted.

20 Gillian Rose documents contemporary geography's ambivalence towards Nature and Mother Earth – an ambivalence which is disquietingly reminiscent of these fascist men's feelings (G. Rose, 1993, Chapter 4 and elsewhere).

afraid of flows: they talked of their fears of waves of Bolshevism which threatened to swallow up eastern Europe; they talked of the red flood which would inundate them; they struggled to dam up Germany against the streams of insurgents pouring in from the east. Their fears of flows also have a bodily location, in their fears of the flows of desire, the flows of (all) bodily fluids. If these men were to dam up, to become a wall against, these flows, then they had to become strong, firm and erect. Once again, the internal map of space, the body and the mind, and external map of space, the body and the social order are resolved one in the other. Through the metaphor of the flood, the kaleidoscope of body, space, subjectivity, politics and the social order is brought into composition. While metaphors of floods, streams, flows, etc., name a terror, they also enable soldiers to find firm ground on which to stand and make a stand.

> The flood is abstract enough to allow processes of extreme diversity to be subsumed under its image. All they need have in common is some transgression of boundaries. Whether the boundaries belong to a country, a body, decency or tradition, their transgression must unearth something that has been forbidden.[21]
>
> (Theweleit, 1977: Volume 1: 232–233)

Soldiers side with prohibition against transgression, with the bound against the unbound, and they want to avoid everything that flows – whether it is paper, promises, politics, or intellectual currents. The fear of being swallowed up by the flood is related directly to, and experienced through, the body: the 'Red Flood' really does threaten to engulf these men. They want to stop the floods: as these rush in, as the torrents rise, as the rivers rage, these men see themselves as the last hope of the German nation. They are the men of steel, a damned and damming army. If the dam bursts, if the masses break through, then soldiers must be prepared to face dissolution – then the blood will really begin to flow (Volume 2: 189). The terrain of their desire, fear, pain and anger is their own bodies, traced through the multiple real, imaginary and symbolic spatialities of body, nation, earth, flood. The political danger of communism and the fear of the power of the masses is etched across the fascist mind, their bodies, geographical space and the social order. Again, Theweleit emphasises the ambivalence of these feelings: as much as the soldier males fear dissolution in the flood, they desire contact with it (Volume 1: 234). They wanted to meet the 'red flood' with the 'white terror', with streams of red blood.

Through the allusive power of metaphor-concepts such as nation, earth, and flood, fascist discourse simultaneously structures and expresses the feelings of men. According to Theweleit, when these metaphors are at their most intense, they reveal that what really threatens to overwhelm their

---

21 Of course, racist discourse in contemporary Britain regularly deploys the image of the flood, of being flooded by 'them' – see Chapter 1 above.

bodies, to transgress their boundaries and dissolve them, is not so much other classes, or other races, or other politics, as the female body, femininity and sex. For Theweleit, fascist discourse is primarily mobilised through its ambivalent relationship with women, femininity and sexuality. Indeed, he argues that 'one of the primary traits of fascists is assigning greater importance to the battle of the sexes than to the class struggle' (Volume 1: 169 and 258). In this understanding (as in Doane's above), invocations of nation, race, the body and so on are intensified and given significance through the re-presentation of the female body.

In fascist discourse, there are two counter-posing versions of woman: on one side, the castrating whore and, on the other, the desexualised mother/sister. For Theweleit, this ambivalence tells both of pre-Oedipal configurations but also of the incest prohibition (Volume 1: 124 and 378 – I will return to this point below). These fascist men were afraid of two types of 'castrating whore': first, prostitutes; and, second, working-class women. The women were supposed to be cold-blooded and capable of perpetrating inhuman acts on men's bodies. Indeed, it was their lack of feelings, their already castrated condition, which enabled them to disarm and castrate men. An exemplary figure in this imaginary was the *flintenweiber*, or rifle-woman, where *flinte* in the nineteenth century referred not only to flintlock guns and to the penis, but also to prostitutes! These women, and women like them, were forbidden: they were nevertheless also objects of desire. Fascist men could not contain their ambivalence towards women: so much so that they rarely wrote about 'good' women, such as wives, fiancées, mothers, sisters, unless they were telling stories where women were in danger or had sacrificed their lives (see Volume 1: 367). For me, this is grimly reminiscent of the ways in which white women's purity was the stake in the Hollywood movies described above.

One popular story told of Sythen Castle, near Haltern, which had been occupied by the Red Army in April and May 1920 (see Volume 1: 84–100). It recounted the depravities of not just the Red Army but also of the 'Red Nurses'. In fascist rhetoric, the white castle had been defiled by the bodies of red men and their whores. Worse, these castrating women had mocked the countess, wearing her clothes and aping her high class, and she is eventually murdered. Through a 'spatial' chain of associations (female body – castle – country), the violation of the countess's body is transmuted into the profanation of the body politic of Germany (Volume 1: 106). In this narrative, the positions of the working-class women and the high-born countess are counter-opposed, but in these Fascist male fantasies either women are castrating or they must have no sex: wherever women become sexual, they must die. Fascist discourse is continually vigilant of incursions across the boundaries between desire and sexuality: where desire is desexualised and sexuality is desensitised. According to Theweleit, the problem for these men is that they cannot escape the mother–son relationship and are trapped into placing women either into this relationship (which is incestuous) or outside it (which is a transgression). For this reason, men must fight everything which reminds them of women, their femininity and their sexuality; instead, they want to wade in blood (Volume 1: 205). Meanwhile, all others that are loathed are feminised and sexualised. If bodies are maps of meaning, identity and power, they are also topographies of desire, fear and loathing.

The production of masculinity, as exemplified by fascist discourse, is unintelligible unless the local interconnections of culture, morality, desire, disgust, pain, nation, geographical space, gender, sex, race, class and so on are traced out. And one site where these articulations are inflected is the soldierly body. This section has drawn on Klaus Theweleit's analysis of fascist discourse in inter-war Germany to show how desire and disgust are mapped simultaneously through psychic forms, bodies, geographic spaces and the social order. The spatialities of the body-ego of the soldiers are imagined and expressed geographically, such that there is no one body, but the performance of bodily practices. Masculinity is being constantly deterritorialised and reterritorialised in the imaginations and politics of fascist men. The important point here is that territorialisations of the body, desire and gender were not just scratched onto the surface of the body, these geographies were deeply felt, experienced and acted on. In this case, the boiling angers of fascist men erupted in violence against people they named as inferior others – political opponents, criminals, homosexuals, recidivists, Jews, and so on – and all in the name of culture.

Theweleit's psychoanalysis of the 'white terror' exposes the psychodynamics of a body-ego-space under threat of perpetual pain, fragmentation and dissolution. Fascist men shored themselves up against their own fears and desire, and they recreated themselves, through murder, terrorism, torture and war – through the dissolution of others. And Theweleit continually warns that this psychodynamics waits in the wings of masculinity – and it is not too much of a leap of faith to suggest that elements of this psychodynamics can be seen in contemporary Serbian and Croatian discourses of 'ethnic cleansing' and the systematic rape of women by soldiers[22] – in, once more, the identification of the nation and the body, and the seething intensity of the desires, fears and violence of fascist body-ego-spaces.

CONCLUSION: URBAN(E) BODIES

The soldiers' hearts and minds, their experiences, the social order and geographic space are articulated by fascist discourse. As we have seen, certain sites become points of capture for intense feelings of desire, fear and disgust: 'nation', 'earth', 'flood', 'woman' and so on, but also 'the city'. Theweleit suggests that through fascist language the soldier's body is connected to larger, external bodies, but primarily the earth and the metropolis. He argues that, for the soldier

---

22 Meanwhile, the British military is currently doing its own insidious moral cartwheels over the place and role of women and homosexuals.

> the evil of that 'witch's cauldron', the big city, also seethes away in his own insides. The political 'order' holding the metropolis in check, for instance, appears to have the same function as the body armour that 'bottles up' his own seething interior. The exploding earth and rebellious metropolis owe their terror primarily to the fact that they embody the potential for – and may violently bring about – an eruption of his own interior. When this interior becomes too powerful, or when the larger, external body (the earth) opens up, or when the metropolis bursts its bounds of order, so that its interior reaches the man's body, the latter is destroyed. In violence and pain, the bodies flow together; their boundaries are exploded: ripping drumhead, eyes wide open, detonating heart.
>
> (Theweleit, 1977: Volume 1: 242)

The city represents the masses and, like the earth, it threatens to engulf soldiers. Nevertheless, the metropolis, like the earth, is not just an external world governed by an external political order: it seethes away on the inside, while the political order is like the soldier's ego-armour;[23] it is a space of desire and fear which is never truly external nor simply internal; the body and the city are mirrored one in the other.

During November 1918, German troops were retreating through Belgium. Their commander, Captain Heydebreck, recalled that

> the demeanour of the troops was exemplary. On my orders, they marched that long march through the streets in tight formation, a corps of merciless avengers, the honour guard of destiny. Not one woman, not even prostitutes, had the audacity to approach them; they would have trampled everything and everyone underfoot. In Prussian goose step, they marched past me through the square in front of the city hall. They halted and stood firm there, a rock amid the surge of the gaping masses.
>
> (1931; cited by Theweleit, 1977: 49)

So, where men fear dissolution into the masses, and fear their engulfment by the female body, the real evil of the city is conjured up in images of sex and gaping bodies. The soldier men stood firm. They were rocks, armoured against the waves of women, prostitutes and the masses that threatened to engulf them. It seems that the terror for them was that they had moved into uncharted territory: women, masses, the city: a territory which was simultaneously psychic, bodily, spatial and social; simultaneously real, imagined and

23 A link might be made here with Lefebvre's suggestion that western culture, as opposed to eastern traditions, produces 'a hard protective shell' around the body and promotes verbalisation (1974: 202). Indeed Lefebvre, like Theweleit, is drawing on Wilhelm Reich to make the argument that individuals need to protect themselves (i.e. armour their egos) using real and imaginary barriers as defence reactions against the 'command', or repressive, culture of the West.

symbolic: but 'real' enough to become the site of violence, torture and murder.

What is being articulated is the relationship between power and 'the body'. What is at stake here, and has been throughout this chapter, is the integrity of body-ego-spaces, a fear of becoming lost in unmapped and unmappable body-ego-spaces and a desire to penetrate and conquer those body-ego-spaces. In the discourses of empire (British and German), psychoanalysis, Hollywood movies and inter-war German fascism, the body, geographic space and the social order are interconnected through their mobilising motifs – whether these are 'the nation', 'the race' or 'the woman'. From this perspective, it is not possible to argue that there is an unmediated relationship between the body and its space: body-ego-spaces are 'territorialised' through the relationship between the individual and the social, between the partial reading of the social map and the personal map-making of the social. This perspective can only be contributed to by integrating some underdeveloped elements: political economies of the body, historical geographies of the body and other spaces of the body, such as 'the public' and 'the private'.

The body-ego-space is territorialised, deterritorialised and reterritorialised – by modalities of identification, by psychic defence mechanisms, by internalised authorities, by intense feelings, by flows of power and meaning. Bodies are made within particular constellations of object relations – the family, the army, the state, the movies, the nation, and so on. These are not, however, passive bodies which simply have a space and are a space: they also make space.[24] They draw their maps of desire, disgust, pleasure, pain, loathing, love. They negotiate their feelings, their place in the world. In their body-ego-spaces, people speak their internal–external border dialogues. Finally, bodies occupy, produce themselves in, make and reproduce themselves in multiple real, imaginary and symbolic spaces, which are never innocent of power and resistance. One site of lived intensity, which we have

---

24 The notion of the body-ego-space is much more in line with arguments that Lefebvre makes while he is drawing on psychoanalysis. For example, Lefebvre states that becoming a human being

> involves reduplication, doubling, repetition at another level of the spatial body; language and imaginary/real spatiality; redundancy and surprise; learning through experience of the natural and social worlds; and the forever-compromised appropriation of a 'reality' which dominates nature by means of abstraction but which is itself dominated by the worst of abstractions, the abstraction of power.
>
> (Lefebvre, 1976: 208)

The problem with his formulation is that, by splitting the natural and the social worlds, Lefebvre presumes that there is a 'nature' to which we can return. Thus, in Lefebvre's schema, the 'total body' does not include the social body. Instead, the idea of body-ego-spaces is intended to disrupt the idea that the social and the natural can be separated in this way. For me, the body has simultaneously real, imagined and symbolic spatialities, which are (all) constituted through experiencing the social body. Once again, bodies are maps of meaning and power, but where these maps subsist in multiple dimensions of space and time.

encountered both in Sirk's *Imitation of Life* and Heydebreck's march past, is the city: a state of mind and body.

7

IN THE CITY
state of mind and body

> Of the several kinds of meaning which may attach to a building or townscape – concrete, functional, emotional, symbolic – it is the symbolic interpretations, rather than intrinsic spatial attributes, which are important in city personality or in the evaluation, choice and attachment to the ideal environment.
>
> (Pocock, 1973: 256)

Drawing on research inspired by Lynch (1960), Pocock believed that the city had a personality only as it was 'determined' by people's evaluation, choice and attachment to the townscape, whether the several kinds of meaning involved in the symbolic interpretation of city personality were concrete, functional, emotional and/or symbolic. While agreeing that the city has no intrinsic meaning grounded solely in its spatial attributes, it was not possible in Pocock's understanding of things to imagine that spatiality was constitutive of the personality of the city: that is, that its concrete, functional, emotional and symbolic meanings were spatially constituted in the city as well as constituting the city as a meaningful (if not ideal) environment. It is Pocock's understanding that the city is constituted within multiple, intersecting webs of significance, but there is no analysis of the ways in which these evaluations are related to relations of power and resistance. It might be said that Pocock deals in expression, but not oppression and not repression. Nevertheless, Pocock has identified some elements which would necessarily contribute to a 'psychoanalysis of (urban) space': from one angle, meaning, interpretation, attachment (libidinal ties) and emotions; from another, the 'personality' of space, buildings, '-scapes', and the concrete, ideal and symbolic aspects of environment (could these be Real, Imaginary and Symbolic spatialities?). Pocock, of course, had no such thing as the psychodynamics of place in mind; nor, it would appear, did Lefebvre, but he too highlights certain possibilities for a psychoanalysis of the city (Lefebvre, 1974).

Without giving Lefebvre's work undue emphasis, I would like to set up this chapter by showing how he draws on psychoanalysis at specific points in his analysis of the production of urban space. Even so, it should not be forgotten that he claimed that psychoanalysis would only be revitalised if it could be proved that the city has an unconscious, or underground, life (1974: 36). Nevertheless, and paradoxically, Lefebvre depends on psychoanalytic precepts, such as the phallus, dream analysis, the mirror, ego and the vicissitudes of erotic life,[1] to advance his analysis of urban space: in particular, of monuments, of streets, of bourgeois homes, of the spatial organisation of ancient cities (Greek and Roman) and of Paris (around 1968). These introductory remarks are intended to excavate the foundations for a 'psychoanalysis of space' but, as I have made clear in Part II, my take on psychoanalysis differs in significant respects from Lefebvre's (remember also the critiques by Gregory, 1995, and Blum and Nast, forthcoming). So, in this chapter, I will not be continually referring back to Lefebvre's work as if it were some kind of baseline. Instead, I will develop his appreciation of the logic of visualisation, phallic space, the body and the erotic life of cities in different directions. In particular, Lefebvre assumes that everyone's experiences of space are the same; instead, I will assume that 'similarities' and 'differences' are socially, personally and spatially constituted. With this caveat in mind, it is possible to gently tease out aspects of Lefebvre's psychoanalytic explication of urban space.

Like Pocock, Lefebvre is interested in the meaning of 'a building'. In his analysis, though, Lefebvre contrasts the banality of buildings (despite their increasingly monumental façades and interiors, see page 223) to the poetry of monuments. For Lefebvre, buildings tell of the 'homogeneous matrix of capitalist space' and, thereby, of the intersection of a particular modality of power with a specific form of political economy (page 227). Although buildings masquerade as monuments, they 'condense' relations of abstract power, property and commodity exchange into particular sites. Buildings are marked by the production of spaces of domination and appropriation. Moreover, buildings 'localise' and 'punctuate' activities in space and time. As such, they 'displace' activities into socially controlled and specially equipped sites,[2] cutting up space and time according to the requirements of capital, which is dominated, maintained and reproduced by the bourgeoisie and the state (see page 227).

Although monuments are contrasted to the production of capitalist space and although they are designated as poetic spaces, they are nevertheless also thoroughly infused with power relations (see also Monk, 1992). This, however, is a different modality of power from the abstract violence of capitalism. Monuments 'speak' of a particular spatial code, which simultaneously commands bodies and orders space (Lefebvre, 1974: 142–143). Monuments make space, urban space, legible (see also the next subsection). So, for example, any 'walker in the city' would be able to recognise the monument as a site of 'intensity', as a work of special significance (page 144). In this way,

---

1 These vicissitudes include psychoanalytic concepts such as repression and other defence mechanisms, repetition, prohibitions, abjection, and so on (see Part II, but especially the Conclusion, above).

2 Lefebvre uses the terms 'condensation' and 'displacement' to describe the double effects of the production of space through buildings – these ideas are taken from Freud's analysis of dreams (see Conclusion to Part II above; and also below).

monumentality produces spaces which, just as it is possible to 'read' them or to see them as the message, are meant to be *lived*. Nevertheless, monuments say what they want to say and, by doing so, they make space incontestable, both by closing off alternative readings and by drawing people into the presumption that the values they represent are shared. Monuments may embody and make visible power relations, but they do so in ways which also tend to mask and/or legitimate and/or naturalise those relationships. For Lefebvre, 'monumental buildings mask the will to power and the arbitrariness of power beneath signs and surfaces which claim to express collective will and collective thought. In the process, such signs and surfaces also manage to conjure away both possibility and time' (page 143).

Lefebvre proceeds to develop his analysis of monuments through two related ideas: first, that monuments make visible and transparent space by providing the means through which they can be 'read'; second, that the monument, as a 'selective' sign and a 'pure' surface, masks the modalities of power which produced it (and space) – whether these power relations are centred on state or masculine brutality (pages 144–147). And it is at this point that Lefebvre's reliance on psychoanalysis begins to show. Monuments, for Lefebvre, have a content: they conceal both a 'phallic realm of (supposed) virility' and a repressive space of surveillance (page 147). While monuments appear to make space transparent and intelligible, they actually produce it as opaque and indecipherable.[3] What remains, though barely concealed, beneath the surface of this analysis is Lefebvre's reworking of psychoanalytic understandings of 'the phallus' and 'the visual' (the separation of the eye and the gaze). From this perspective, monuments re-present 'the phallus': they both make visible and 'mirror' back to the 'walker in the street' their place in the world, geographically, historically and socially; they reproduce repressive spaces which, while ostensibly acting as celebrations of events and people, have both feet in terror and violence; and they repeat not just people's experiences of their bodies and their relations to others but also modalities of power. Symbols on the royal road to the 'unconscious' of urban life may not be that hard to find; indeed, it is possible to analyse urban spaces as if they had been dreamt.

Lefebvre argues that two 'primary processes' operate in social space, the space of social practices as produced by architects, as embodied in monuments. These 'primary processes' are displacement and condensation, 'as described by certain psychoanalysts' (page 225). Lefebvre draws on these ideas at a significant point in his analysis of spatial architectonics; that is, when he wants to be able to move in an analytic space which simultaneously incorporates the body, social space and language. Freud's analysis of dreams and Lacan's structural linguistics are recombined so that Lefebvre can talk about the ways in which monumental spaces both displace and condense meaning and thereby provide 'the metaphorical and quasi-metaphysical

3 On these ideas of transparent and opaque space, see Soja, 1989.

underpinning of a society' (such a method has been called depth hermeneutics).[4] From this perspective, a monumental space, such as the Taj Mahal, displaces meaning by making the spatially-circumscribed site stand for something universal – such as love or death or holiness. Moreover, in this case, the Taj Mahal condenses meaning by substituting one set of meanings for another set of meanings; in this instance, by substituting a sign of royal authority with a celebration of love, death and religion. In this substitution, the predominance of one set of meanings (Taj Mahal = power of love) masks the possibility of reading out other sets of meanings (Taj Mahal = royal power, arbitrary power, brutality, and so on). This is why it has been described as 'a tear on the face of eternity',[5] rather than as another skull in the history of oppression and terror.

Monuments are, for Lefebvre, significant co-ordinates (sites of intensity, points of capture) in matrices of meaning and power, where these are simultaneously cartographies of social, historical and geographical space, and where these cartographies are written through, and read through, the body, subjectivity and language. Freudian and Lacanian tenets enable an analysis of social space, in general, and of monumental space, in particular, which charts these spaces by placing them within chains of association and substitution which, horizontally and vertically, link up to masked aspects of these spaces. In Lefebvre's analysis, monuments are not just spaces of the body, subjectivity and language, but also grids of meaning and power, which are complicit in the control and manipulation of simultaneously real and metaphorical space, where for example chairs become thrones, buildings become monuments, and so on. Moreover, a psychoanalysis of space-as-dream can reveal the unconscious processes involved in the production of space: that is, the condensations, the displacements and the primary and secondary revisions involved in the constitution of urban spaces.

So far, Lefebvre has used psychoanalytic understandings of the phallus, the visual, repression, repetition and dreams to interpret the production of urban space: specifically, the predominance of a visual-spatial regime and the use of monuments in the production of Parisian space (see pages 312 and 386 respectively). He uses another psychoanalytic co-ordinate, the Ego, to map the space of interaction between the house, the family and genital sexuality (see pages 121 and 223). Following Gaston Bachelard (1957), Lefebvre argues that the city and the House have the qualities of a memory, of a persistent dream. Moreover, the House, like the city, is a multitude of spaces: 'from cellar to attic, from foundations to roof, it has a density at once dreamy and rational, earthly and celestial' (page 121). Lefebvre continues along this chain of spatial analogies to a point where he suggests that 'the relationship

4 For a review, see Pile, 1990.

5 These are the words of the Indian classical poet Tagore. And my confession is that this is the way I experienced the Taj Mahal when I went there recently. I quickly forgot the twenty thousand who sweated and died over twenty-one years to build the thing. I took some beautiful slides . . .

between Home and Ego ... borders on identity'. Lefebvre intends to say more than simply that the Home and the Ego are spaces of identity. The Home and the Ego are almost identical spaces, because both are, following Bachelard, 'a secret and directly experienced space' (page 121). I do not think it unreasonable to suggest that what Lefebvre has in mind here is a Freudian sense of bodily sensations being experienced through 'Home' and 'Ego'. Indeed, Lefebvre's spatial understanding of the Ego, as Freud's, enables him to mix and match seemingly separate scales of analysis (similarly, see page 311). However, such an interpretation tends to underestimate the intermediaries between external and internal space, between the Home and the Ego (see also Chapter 3 above).

Later in his book, Lefebvre writes another passage which glides quickly from place to place. This time the chain of associations is not provided by the Ego, but by erotic life. In his analysis, the urban fabric 'suffers' from emotional intensity: whether it is expressed in street violence, in the investments in monumental space or in the homes of the moneyed classes. Two further aspects of Lefebvre's psychoanalysis of urban space are thereby revealed: first, is his belief that the bourgeois family and home is organised around genital sexuality (see, for example, page 232); and second, is his conviction that bourgeois space expresses a repressed erotic life (page 315).[6] In both cases, Lefebvre's examples are urban, although his analysis is not (really) confined to bourgeois life and space. He does not, however, develop his argument that the bourgeois family is organised through genital sexuality and, therefore, that bourgeois homes and house design express genital sexuality in any detail. The point Lefebvre makes, quickly, is that city planning and house construction is imagined in terms of a family which (a) stays together and (b) reproduces.[7] While such an argument does not require psychoanalysis, Lefebvre seems to share Freud's suspicion of the 'normal' family and 'normal' genital sexuality.

Lefebvre does not confine his analysis of erotic life to the bourgeois family.[8] He also sets up his discussion of the differences between Greek and Roman cities by, first, talking about the ways in which 'horizontal space symbolises submission, vertical space power, and subterranean space death' and, second, the ways in which absolute space is subject to the mental mechanisms of identification and imitation (see page 236). Greek and Roman

---

6 See also Blum and Nast (forthcoming) on Lefebvre's analysis of the *imago mundi*. They argue that there are similarities between the way in which Lefebvre discusses the role of the *imago mundi* in the constitution of subjectivity and Lacan's understanding of the mirror stage. They argue, further, that both Lefebvre and Lacan reveal in their analysis a heterosexual presumption.

7 Feminists have made similar observations (see Matrix, 1984; Little, Peake and Richardson, 1988).

8 See also Lefebvre's discussion of the *mundus* and Paris's red light areas (pages 242 and 320 respectively – see also the Introduction to Part III above). In his analysis, Lefebvre draws on psychoanalysis to show how dirt, the body, sex and dark places are transcoded and abjected (see also Chapter 6 above).

cities, for Lefebvre, speak not only their inseparably social and mental constitution but also their erotic life. For Lefebvre, the Greek city state married politics and religion, on the one hand, and a metaphysical and mathematical rationality, on the other. From this perspective, Greek cities were segregated into secular and sacred sites, which not only tell of the dominant social order but also of the way it was 'governed by a good many prohibitions' (page 240). Indeed, the vicissitudes of political and religious power and mathematical and metaphysical rationality formed a kind of real 'unconscious' of the Greek city for Lefebvre (page 241).[9]

In contrast, the Roman city answered the demands of the state, the judiciary and 'the bodies and minds of free – and rich – citizens' (page 239). While Athens was subject to the law of the Logos/Cosmos, Rome was built on the imperatives of a slave mode of production. Nevertheless, Roman law is governed by what Lefebvre calls, in a Lacanian moment, 'the figure of the Father' (page 243). The space of power in Rome was constituted through a specific relationship between property and patrimony. It would seem to be logical to suggest that Rome was a map not just of Roman law but also of 'the Law of the Father': by extension, Rome is the paternal metaphor (in Lacanian terms). The Roman city provided a simultaneously social, personal and spatial matrix of meaning, identity and power which not only enabled people – as citizens, women, slaves, servants, children, soldiers, and so on – to find their proper place, but also confined them – as men or women, free or unfree – to that place. Remarkably, Lefebvre observes that what has been lost is a maternal principle.[10]

The idea that the maternal has been lost is not a casual throwaway line (and, I should add, it resonates with psychoanalytic notions that children must 'lose' their love for the mother). Despite rejecting the idea that the city can be understood through an Oedipal grid (page 248), Lefebvre nevertheless argues that both Greek and Roman cities are constituted around a masculine principle (is that a phallic, or Oedipal, principle?).[11] The Greek city was organised on a principle of 'manliness', according to Lefebvre (page 249), while the Roman city was centred on 'masculinity'. While the Greek city was

---

9 Lefebvre puts the term unconscious in scare quotes, but I would suggest that this is to emphasise that he is drawing on a Freudian (rather than everyday) understanding: i.e. because Lefebvre is using it to describe contradictory ideas in tension through space which is both there and hidden (see pages 240–241).

10 A similar argument is made by Mumford (1961). He argues that Neolithic agriculture was organised around the 'home and mother' and that cities could only develop where women's husbandry produced surplus capable of supporting an urban population, which was dominated by men and trade (see pages 19–24, 36).

11 Lewis Mumford (1961: 22) argues that the production of space was organised around the body (see also Sennett, 1994): for example, bowls were supposedly based on the breasts of goddesses while male statues were given huge penises. These symbolisations gave rise to specific architectural forms: columns, domes, monuments, and so on. From this perspective, there is a maternal (vaginal) as well as a phallic formant of space; whereas, for Lefebvre (as in classical psychoanalysis), the female genital organ is hidden (and represents the world), while the phallus forces its way into view and comes to represent (male) privilege and

built on notions of justice grounded in the discrimination between challenge and defiance, that is in manly conduct, the Roman city was founded on submission to the Law, where military and power élites commanded. If we are not bound to a rigid Oedipal or phallic interpretation of these urban spaces,[12] then it is possible to suggest a psychoanalysis of space which correlates sexuality, geographic space and power with the body, meaning and (real, imaginary, symbolic) spatialities. In this case, it can be argued that Lefebvre shows how different masculine principles are involved in the production of specific bodily, personal, social and urban spaces. Hence, for example, the importance of particular kinds of monuments and architecture in the production of cityscapes. Moreover, this analysis demonstrates that Athens and the Greek body politic and Rome and the Roman body politic are simultaneously empires of subjectivity, space and power.

In Lefebvre's description of cityscapes, I have shown that he selectively draws on certain psychoanalytic tenets and co-ordinates: the phallus, dreams, mirrors, repetition, identification, Ego, genital sexuality, sexual differentiation, the Law of the Father and repression and erotic life. From the discussion of psychoanalysis in Part II, it can be seen that there are problems with some of these ideas. Nevertheless, this discussion demonstrates that the idea of psychoanalysing the city is not quite so fanciful as it might at first seem.[13] It also, I believe, confirms that the resources I suggested might be used in a psychoanalysis of space can be selectively applied to the production of the urban as a state of mind and body.[14] This chapter takes certain of the themes raised already and develops them from this different understanding of a psychoanalysis of space. First, I will deal with a psychoanalysis of the image of the city through a discussion of Kevin Lynch's *The Image of the City* (1960), partly in order to raise questions about subjectivity and space, and partly in order to talk about place as simultaneously real, imagined and symbolic.[15]

(masculine) power (see page 262). This view, I believe, both challenges and supplements Lefebvre's urban history: challenges by replacing the feminine, supplements by recognising the phallic.

12 That is, following the critiques of Freud, Lacan and Lefebvre established in Part II above.

13 Not that it should seem fanciful: as far back as 1961, Lewis Mumford suggested that psychoanalysis has been able to bring to light the tawdry meanings of both symbols and architectural, in particular inner or secret, space (1961: 22–23). In this frame, psychoanalysis interprets not only the sexual referents of space but also space as security, nurture, container, enclosure, welcome, protection, sanitation, sanity, civilisation and so on, from the house, the oven, the bedroom, to the village, to the city, the street, the cloister, the atrium, the wall – to the city as body, the city as erotic order.

14 I will not repeat them again, see the Conclusion to Part II, the Introduction to Part III and the introduction to Chapter 6 above. I will however return to these ideas in Chapter 8 below.

15 For, as Lefebvre says, 'a discourse on space implies a truth of space, and this must derive not from a location within space, but rather from a place imaginary and real – and hence 'surreal', yet concrete. And yes – conceptual also' (1974: 251).

IMAGINING THE CITY: IMAGE, IDENTITY AND URBAN SPACES

The Image of the City (1960), as noted in Chapter 2 above, was a source of inspiration for behavioural geographers throughout the 1960s and 1970s. They picked up on the 'mental maps' which Lynch's researches had solicited from the residents of three North American cities (and I hope to reinvigorate this project by reinstalling aspects of psychic reality which did not appear in Lynch's analysis). Although Lynch admitted that his interviewees were hardly representative, geographers were excited by the idea that the environment was acted on, not in terms of its physical reality, but in terms of the mental pictures that people had built up of that environment.[16] People did not live in the city as such, but inside the mental picture that they had built up of the city. In their excitement, geographers have – not unreasonably – concentrated on certain aspects of Lynch's arguments (the mappable ones). Nevertheless, in his first sentence, Lynch says that what he is interested in is 'the look of cities' (page 1). Of course, Lynch is not secretly invoking Lacan, or Freud, or Lefebvre. Nevertheless, I will. Behavioural geographers paid much less attention to the fact that Lynch's book is a piece of rhetoric about the 'good' city, which relies on a particular understanding of human perception.[17] By examining the look of the city, notions of quality and the body, it is possible to mark out some important sights/sites for the analysis of the city. I will deal with these in this order: the body, the quality of the environment and the look of the city.

Lynch uses a wealth of information, derived essentially from anthropology and psychology, to describe the tremendous adaptability of human beings to their environments, but also their ability to give meaning to those environments and to change them. In Appendix A (pages 123–139), Lynch relates how different peoples survive in extremely hostile or disorientating environments, which in some cases are apparently featureless – such as deserts, jungles, ice fields, the oceans. The evidence demonstrates, he argues, that there is a two-way process between the observer and the observed in the building of 'the image' of the environment (pages 6 and 131). While there is no direct link between any physical environment and the systems of thought that people used to get around and make use of (urban) space, Lynch suggests that there are common features: first, these systems of orientation are both emotional and practical, they make people feel safe and enable their survival; second, they are generalised, coherent and selective; and, third, they relate to the body – not just what is seen, heard and felt, but also what is needed, imagined and significant. Thus, 'the symbolic organisation of the landscape

---

16 In psychoanalytic terms, these mental pictures might be seen as either 'psychic reality' or 'object relationships' or both.

17 Lynch's argument – about 'good' environments and about the nature of the body and human perception, especially senses of direction, space and time – is underpinned, it should be noted, by evidence from 'primitive' races (for a related critique, see Chapter 6 above).

may help to assuage fear, to establish an emotionally safe relationship between men and their total environment' (page 127). The total environment is experienced through the total body, which – like Lowenthal's (1961) – sees colours, shapes, motion, shades and so on, but also smells, hears, touches and moves, and possibly feels gravity, electricity and magnetism (like other animals) (1960: 3). The body is a 'total reality' and so is the external physical world, unlike the symbolic organisation of the landscape, but people act on and in this symbolic world. Thus,

> in the process of way-finding, the strategic link is the environmental image, the generalized mental picture of the exterior physical world that is held by the individual. This image is the product both of immediate sensation and of the memory of past experience, and it is used to interpret information and to guide action. The need to recognise and pattern our surroundings is so crucial, and has such long roots in the past, that this image has wide practical and emotional importance to the individual.
>
> (Lynch, 1960: 4)

Quickly, it can be noted that these ideas could be translated into the notions of Real, Imaginary and Symbolic spatialities without doing any harm to Lynch's argument. It can also be noted that Lynch's 'environmental image' sounds a little like Freud's body-ego, or even a body-ego-space. Lynch's argument, however, concerns the production of cityscapes which have particular imaged qualities: coherence, safety – and legibility. In his analysis, Lynch argues that legible environments not only provide security but also add to 'the potential depth and intensity of human experience' (page 5). Since Lynch wants to produce an urban space which is shaped for 'sensuous enjoyment', which is 'beautiful and delightful' (page 2), he argues that this city must be legible: that is, it must be able to 'be grasped as a related pattern of recognisable symbols' (page 3). Thus, an integrated legible environment can 'furnish the raw material for the symbols and collective memories of group communication' (page 4). The benefits of a legible environment are not just collective: a good cityscape will let people know where they are – and this knowledge will provide the background 'for individual growth' (page 4). While people may enjoy crooked streets or a Hall of Mirrors, Lynch argues that they are basically most fearful of being lost (hence the use of the phrase to describe multiple terrors). The good cityscape, then, prevents loss and, thereby, becomes the potential repository of actions, memories and expressive meaning (page 5). It can be noted, at this point, that Lynch's sense of the urban as a potential site for personal meaning and collective communication contrasts with, but can also add to, Lefebvre's vision of the city as a system of signs representing relations of power.

In his evocative introductory remarks, Lynch builds a dense, ambivalent, but curiously static, map of the city – it is stage, spectacle, pleasure, delight, beauty, ugliness, smell, noise, fear (of loss), symbol, knowledge, route, memory, purpose, geography and, above all, text. Nevertheless, in the two-way relationship between observer and observed, Lynch stresses the production of a rational space which is capable of conscious manipulation (i.e. reading and writing the urban script), and he presumes that this 'imageable' urban space should be produced by urban designers (i.e. urban planners and architects). In his analysis of the look of the city, people locate and orientate themselves using five basic elements: path, landmark, edge, node (i.e. 'a condensation of some use or physical character', page 47) and district (involving a sense of 'inside–outside'). Looking at the city means envisioning a selection and recombination of these features, where each of these components has three aspects: identity, structure and meaning. Identity means being able to identify a feature (i.e. identity means the object's difference from other objects), while structure means the object's spatial relationship or pattern in relation to others. Meanwhile, the object has meaning for the observer, whether this meaning is practical or emotional.

Despite Lynch's avowed and repeated aim of building cities for enjoyment, he retreats from analysing the meaning of the basic elements of the cityscape, concentrating instead on identity (difference) and structure (spatial relationships). There are two observations I would like to make: the first is that Lynch retreats from an analysis of pleasure, and also of fear, while overlooking completely relations of power; the second is that he actually pulls out only two basic elements of the city – the landmark and the street.[18] At this point, Lynch's work uncannily starts to reinforce Lefebvre's analysis of 'monuments' and 'streets' as producing a legible cityscape which 'condenses' and 'localises' both meaning and abstract power relations. Replacing pleasure, fear, the symbolic and signification in the interpretation of landmarks and streets would be one way to psychoanalyse the city. If this seems like stretching Lynch too far, then it might be salutary to think about his sense of Manhattan (frequently cited as an 'imageable' city): 'The image of the Manhattan skyline may stand for vitality, power, decadence, mystery, congestion, greatness, or what you will, but in each case that sharp picture crystallizes and reinforces the meaning' (1960: 8–9).

Lynch, from another perspective, is so impressed with the Manhattan skyline because of its phallic power – his senses of its vitality, mystery, greatness (terms he never otherwise uses) are erected on (the height of) Manhattan's towers. If Lynch's work only unconsciously acknowledges a phallic formant of space, it explicitly concentrates on the visual formant (apparently, without having read either Lefebvre or Lacan). For Lynch, the look of the city is practical and emotional: the skyline 'speaks' the city, it also 'speaks' the observer – the moment he chooses to talk about power and decadence is when he is confronted by Manhattan's phallic arrogance, while the streets below (like anywhere else) are congested, drab, dirty and noisy. While power – phallus or capital – is an invisible presence which provides a

18 That is, paths and edges are 'streets' and edges, nodes and district are simply 'landmarks' at differing scales.

sharp picture which reinforces meaning, the visual elements – each with their 'façade' – of the landmark and the street are the sites of intensity of everyday human experience. From 'the tower' to 'the street', a psychoanalysis of the city can also pick up on Lynch's restrained – but resolutely dynamic and ambivalent – pleasures and fears: enjoyment, sensuality, dirt, congestion, beauty, ugliness, depth, mystery – in a cartography of subjectivity, meaning and power.

TOWERING: MEAGHAN MORRIS AND THE LIMITS OF PHALLIC ARROGANCE

From a Lefebvrian or psychoanalytic perspective, it can be suggested that the Manhattan skyline is not just about an alliance between a phallic formant of space and abstract (capitalist) power relations, but also about a bourgeois coupling of 'Ego' and 'Phallus'. Space is produced under the tyranny of three intersecting, aligned lines of power: masculinity, the bourgeois family and capitalism. So, in concert, these powers produce the rhythms of New York, New York.

A few years back, Nelson's Column, in London's Trafalgar Square, was cleaned. They covered it in plastic sheeting, partly to prevent the high pressure water jets from spraying surrounding passers by. Meanwhile, nearby, a postcard was being sold of the Column under its protective sheath: the caption read 'Nelson's Condom'.

Monuments are phallic because they are erect, tall and associated with power; while high buildings exhibit corporate power by being erect, tall and associated with wealth: but (and this is a big but) the phallic can be brought down to size by associating it with its fleshly correlate. There is something obvious and unacceptable about all this phallic exhibitionism, according to Meaghan Morris (1992). She interprets two rather different social spectacles of phallic verticality – the use of King Kong in an advertising campaign for a Sydney tower and the climbing of Sydney Tower by Chris Hilton, who was described by the press as the Human Fly.[19] She finds that there is a shift from one spectacle to the other, between the phallus/penis elision of corporate discourse and the production of a face/faciality relation in the radical political tactics of the Human Fly.[20] Although Morris maintains a certain commitment to postmodernism and a suspicion of psychoanalysis which runs against the grain of my argument,[21] I believe that there is much to be gained

---

19 Unfortunately, I will only be able to refer to certain aspects of these 'spectacles' (and other fascinating elements of Morris's analysis) in passing.

20 Her understanding of the term 'tactic' (and 'strategy') is taken from Michel de Certeau (1984). The tactic is 'a mode of action determined by *not* having a place of one's own', while the strategy is 'a mode of action specific to regimes of *place*' (Morris, 1992: 32; see de Certeau, 1984: xix–xx, 34–39).

21 Even so, she does draw heavily on the work of Deleuze and Guattari (1987), noting their 'para-sitic' relation to psychoanalytic discourse (Morris, 1992: 5).

from her line of questioning: in particular, her refusal to set the Phallus up as an all-powerful place of privilege and power, and her commensurate offsetting of (the politics of) bi-polar opposites.

In the first place, she cites an Australian developer as saying, 'The tower is not an ego thing. You don't spend a billion dollars on ego' (John Bond, 1987; cited by Morris, 1992: 6).

Morris initially notes that Australian corporate culture in the late 1980s might be different from that exhibited in the United States where, if you had a billion dollars to spend on a building, you would do it big, as did Donald Trump. But she is much more interested in the question of what it is that John Bond is refusing: that is, why should he bother to deny that the tower *is* an ego thing? What is an ego thing? She argues that John Bond is irritated by the popular view that the size of tower is inversely proportional to the size of penis. The display of corporate power is mocked as an exhibitionism, grounded in a personality defect: 'having it is one thing, flaunting it another' (page 9). John Bond's aside explains that his investment is in hard bricks and mortar, not in armouring his ego inadequacies. The height of the tower, Bond is saying, has nothing to do with his body-ego-space. It seems, to Morris, that the casual identification of status symbols as penis extensions has some efficacy (if only because it needled Bond); but, on the other hand, there is something disquieting about this – it is too easy to unmask and mock the supposedly veiled phallus.

> Popular mockery of the tower form as a (male) 'ego' thing involves some ambiguities. It is *vaguely* antiphallic: while it assumes that a tower is a 'phallic symbol' (in this code, a penis extension), the force of the insult is that someone's *ego* is also a penis extension: in the vernacular, 'that bloke thinks with his dick'. (This, presumably, is partly what Bond was denying on his own behalf, rather than that desire can be invested in making a lot of money.) But at the same time, a controlled and controlling 'masculinity' is reaffirmed as the norm of public conduct. An ego thing is shameful because too prominent, too visible to others; one is *caught*, or exposed, at 'doing' an ego thing; it's a form of unseemly display, and thus a sign of effeteness (why else should Bond deny it?).
>
> (Morris, 1992: 8)

The tower, in Morris's analysis, must be something other than a phallic space, if it is so easy to expose it as an overt, unseemly penile display (following Gallop, 1982): that is, Nelson's Column isn't quite so awe-inspiring when it is transformed into a profane prophylactic, so its power must also subsist in other spaces. In her estimation, jokes along the lines of Lefebvre's elision of Ego and Phallus have some force because they shrivel the prestigious monument or skyscraper to the hu*man* scale of social pretentiousness and fleshly defects. Nevertheless, according to Morris, these phallic jokes only serve to mask the brutal, vicious and abstract power of corporate capital by reducing the urban landscape to a singular, psychosexual cause (page 9). Citing Harvey (1985) and Huxtable (1984), she reminds us that skyscrapers are constructed within the technological, political and economic circumstances of the urbanisation of (speculative, spectacular) capital, that building towers has material, detrimental consequences for the poor, the marginalised and the excluded, and that the forces making the urban landscape are somewhat arbitrary, internally contradictory and repeatedly contested (see also Domosh, 1992).

There is some doubt, then, as to the phallic status of tall erections, such as monuments and skyscrapers. Moreover, for Lefebvre and Lynch, the modern city is a master-planned monument to the power of an invested idea: *la ville radieuse*, Paris, New York. However, in the entrepreneurial city of the late 1980s, the spatial stories are both more, and less, about the elision of Ego, Phallus and abstract power. Instead, for Morris, stories of the city are not only about the global mobility of capital investment, but also about face, faciality and façade. As for Deleuze and Guattari (1972, 1987), flows of capital and desire are etched on the face of the earth, continually deterritorialising and reterritorialising the body and the city. In this analysis, the shift from phallus/penis to face/faciality is significant because it marks the deterritorialisation of one masculine space in favour of another: thus, the substitution of the body of the White Man by the Face is allegorically read as the substitution *of* the brutal body politic of urbanised capital *by* the cityscape as a collection of postcard scenes (and this is why tall buildings have observation decks and revolving restaurants – to construct the city as a framed panorama).[22] From this perspective, the Manhattan skyline becomes the perpetually acceptable face of capitalism. And Chris Hilton's climb up the Sydney Tower is an act of resistance because, by becoming the human fly on the face of the tower, he disrupted modalities of power which produce the segregation of public space and private space, of Culture and Nature, of adventure and home – where, in this form, the first terms are sexed male and the latter are feminised.

If the only possibility for the Tower is that it be the Phallus, then there is no space for the subversion of the Tower's place as a fixed co-ordinate in an intensifying grid of meaning and power. In Morris's view, the Tower can be cut down to human size – from Phallus to penis, through a Face to face encounter (a climb). The Tower-Phallus is quickly unveiled and produced as an other space. For Morris, this other space is tactical (an action not in its 'proper' place, the denial of a prohibition); is a spectacle (by showing himself up, the climber reveals the extent to which the tower is a male ego thing); and is the production of new meanings through resymbolisation (that is, by associating the Sydney Tower with the shared imaginary Human Fly). What she resists is the reduction of space to a single, psychosexual object – the

22 It can be added that this is also why high buildings are made of metal skeletons, where the face/façade can be replaced, as desired. The tall building is not just a phallus, or a prosthetic dildo, but also a cyborg, awaiting plastic surgery. It is a very Modern Thing.

Phallus – by evoking other aspects of the object – the Face. This move adds to a psychoanalysis of space, rather than subverts it, not just by situating the production of space within constellations of object relations and by teasing out the specific modalities of dialectical identifications, but also by turning on other aspects of the object in its real, imaginary and symbolic spatialities.

From this perspective, it can be concluded that jokes and façades, by transforming the Tower into a fleshly penis, actually serve to veil its function as a Phallus (a space which gives meaning, which produces a legible urban space). Nevertheless, while the Tower is Phallic, there are limits to its arrogance: it is neither all-powerful, nor all-seeing, nor all-encompassing. The Tower is both more and less than the Phallus. It is rather a space produced as a performance, as a masquerade; and it is a fetish that just thinks it is more important than other spatialised fetishes.[23] The Tower has no essential identity and is not more significant, but it nevertheless 'behaves' as if it does and is, or rather people behave as if it does and is – which is why the Human Fly's climb is radical, because it calls into question the identity and grandeur of the Tower.

The Tower is the production of a space as a script, a spectacle, a sex aid, a vision, a sight and a site, but none of this means that it has an identity. Its identity is constantly shifting – Phallus, Ego, Face – and it thereby remains a veil, a chameleon, a hybrid of non-essences, deterritorialised and reterritorialised by flows of desire and power. This interpretation helps explain why, for Lynch, the Manhattan skyline may stand for so many things, yet paradoxically also crystallise and reinforce meaning (see above). The Manhattan skyline is open to psychoanalytic readings precisely because it cloaks, exhibits and performs desire and power, through a political economy of dominant visual and phallic spatialities. These spatialities primarily mark the real, imagined and symbolic Tower as a presence, whether this is seen as phallus, ego or face. The Tower is, thereby, produced as a space – and produces a space – visible both for the dominant regime of the look and for those who walk the streets. We should also note, however, that this space is fixed by, and fixes, the ambivalences of absence and presence, distance and proximity, spectacle and modesty, meaning and power. We might begin to unpack other urban psychodynamics by considering the experience of one man up a tower, looking down at the streets (following on from Morris, 1992: 13). And following the paths he takes.

STREET WALKING: THE FLÂNEUR, THE PROSTITUTE AND PROPER SPACES

Perhaps curiously, de Certeau begins his chapter on the spatial practice of 'walking in the city', with a description of his experience of looking at the New York skyline from the observation deck of the World Trade Centre:

23 Following the arguments made by Gallop, 1982 and de Lauretis, 1994 (see Chapter 5 above).

> Seeing Manhattan from the 110th floor of the World Trade Center. Beneath the haze stirred up by the winds, the urban island, a sea in the middle of the sea, lifts up the skyscrapers over Wall Street, sinks down in Greenwich, then rises again to the crests of Midtown, quietly passes over Central Park and finally undulates off into the distance beyond Harlem. A wave of verticals. Its agitation is momentarily arrested by vision. The gigantic mass is immobilized before the eyes.
>
> (de Certeau 1984: 91).

De Certeau argues that the view from the tower turns the city into a text and that this perspective blocks out the view of the pedestrian, the person in the street. This view from above creates a city scene (sea and seen) that celebrates and demonstrates its dynamism, the excesses of money and power, and its energy and vitality (much the same city, indeed, which Lynch described when seeing it in panorama). De Certeau continues:

> to be lifted to the summit of the World Trade Center is to be lifted out of the city's grasp. One's body is no longer clasped by the streets that turn and return it according to an anonymous law; nor is it possessed, whether player or played, by the rumble of so many differences and by the nervousness of New York traffic.
>
> (1984: 92)

Carried away from the streets, from the crowds and from the ordinary man, the pedestrian is turned into a particular kind of spectator: a voyeur. There are two points which I would like to make here: first, de Certeau builds his analysis out of a reading of the 'text' of the city, deploying an analogy between walking and narration; and, second, throughout his analysis, there is an implicit – and sometimes explicit – sense of the erotics of knowledge. Because a discussion of this contributes to an understanding of the psychodynamics of place, I would like to develop these issues in detail. In this subsection, I will begin to trace out the intersecting erotics of knowledge, vision and street walking. I will be arguing that power and desire are traced through historically and geographically located visual regimes which subsist in the spatial practices of the city – whether the viewer is in the tower or the streets. There is, as Lefebvre suspects, an 'underground' erotic life in the city – which, I suspect, does not depend on how close to ground level you are.

De Certeau argues that walking narrates interests and desires that are neither determined nor captured by the system of signification used to codify them, but within which spatial practices develop. From the tower, de Certeau saw a city immobilised before his eyes. He believes that he has been forced both into a place, and into becoming the subject of a particular desire, from where space is made to seem readable. The production of legible space has fixed co-ordinates: it is mappable and provides a hard background to knowledge. Through a process of naming, the city 'provides a way of conceiving and constructing space on the basis of a finite number of stable, isolatable, and interconnected properties' (page 93). The City organises space not simply by naming – and thereby creating an identity which presumes an essence but has no essence in advance of being named – and also by citation (page 188) which, by reference to other spaces, gives an apparent shared meaning to space, as if everyone believed it.[24] From this perspective, The City is a means of organising space into its proper place – it cuts up, parcels and names houses, streets, neighbourhoods, blocks, and so on.

This argument is reminiscent of Lefebvre's understanding of monuments and Lynch's imageability, but what de Certeau stresses (in contrast to Lefebvre, but extending Lynch) is the innumerable ways in which walking in the streets mobilises other subtle, stubborn, embodied, resistant meanings. The streets become haunted by the ghosts of other stories (see also de Certeau, 1984: 186). The city becomes a ghost town of memories without a language to articulate them because walking is a transient and evanescent practice. The proper spaces created for the city by the view from above – whether embodied in the visual regimes of the panoptic gaze or cartography – are interrupted, resignified and torn by the everyday practices of moving by foot. Walkers are involved in the production of an unmappable space which cannot be seen from above (metaphorically from the 110th floor). In this unmappable space, walkers ceaselessly move around in 'spaces of darkness and trickery', ever-refashioned by a 'combination of manipulation and enjoyment' (page 18).

> The ordinary practitioners of the city live 'down below', below the thresholds at which visibility begins ... they are walkers ... where bodies follow the thicks and thins of an urban 'text' they write without being able to read it. These practitioners make use of spaces that cannot be seen; their knowledge of them is as blind as that of lovers in each other's arms.
>
> (1984: 93)

I will return to the lovers analogy, but for the time being it is important to note the relationship between reading and writing. The bodies of walkers write, but cannot read what they write. De Certeau implicitly (and elsewhere explicitly) is drawing on Lacanian notions of language and the Real to suggest that walking involves (what I call) Real, Imaginary and Symbolic spatialities. The Real city is lost and hidden, in this sense walking always involves a lack of place (not being somewhere). Meanwhile, this Real spatiality, while determinant, is unreadable – that is, unconscious. Just as language is the movement of tropes (from Freud's dream analysis and Lacan's structural linguistics), so is walking. Just as language seems to create, convey and elucidate meaning but also fails (following Lacan's notions of

---

24 Superficially, this view seems to parallel Judith Butler's analysis of identity (see Butler, 1990). This 'coincidence' becomes less surprising when it is realised that both Butler and de Certeau rely heavily on Foucault and psychoanalysis.

lack and Real, rather than the significance of the phallus), so walking enunciates and appropriates the space of the city but also sets up its own prohibitions and actualises only some of the possibilities of moving around the city.

'From above' and 'from down below' are two opposing positions in de Certeau's analysis. Metaphorically, the 110th floor represents the godlike eye of the powerful, the voyeur, while the street is the skilled vision of the ordinary practitioner. In this way, de Certeau creates a 'mythological city' out of the combination of manipulation and enjoyment which is beyond the ability of power to see, conceive and control: 'Beneath the discourses that ideologise the city, the ruses and combinations of powers that have no readable identity proliferate; without points where one can take hold of them, without rational transparency, they are impossible to administer' (1984: 95).

The City has an unconscious life which, we might say, carries out a guerrilla warfare with attempts to repress it: in other words, administrative rationality continually struggles to impose an order on people's everyday urban spatial practices, but must always fail. There are psychoanalytic analogies in this argument: the unconscious and the id ('The City') are beyond the control of the conscious and the ego ('Administrative Rationality'); or we cannot rationalise our dreams, only revise them after the event. The City, the view from above and walking are fixated – like the lovers who are blind – with erotics, closeness and a way of seeing. The analogy between the interconnected triads of loving-blindness-knowledge, walking-writing-knowledge and loving-walking-space is neither casual nor fanciful. De Certeau has a particular story in mind and, somewhat surprisingly, he ends his discussion of walking on a psychoanalytic story: the Fort/Da game (see Chapter 5 above). Place, he concludes, is

> the repetition ... of a decisive and originary experience, that of the child's differentiation from the mother's body. It is through that experience that the possibility of space and of a localization (a 'not everything') of the subject is inaugurated ... The childhood experience that determines spatial practices later develops its effects, proliferates, floods private and public spaces, undoes their readable surfaces, and creates within the planned city a 'metaphorical' or mobile city...
>
> (1984: 109)

This ending is both bizarre (we are not led to expect this conclusion) and troubling (space is once more feminised). Walking, thus far, has been analysed as a fleeting, though ineradicable, space of resistance, now suddenly spatial practices are determined by the repetition of childhood experiences of loss of the mother (which, by magic, proliferates spaces and undoes surfaces). This move is less surprising, however, when de Certeau's emphasis on the relationship between manipulation and enjoyment is taken into account: the Fort/Da game speaks, for de Certeau, of the child's manipulation and enjoyment of objects in a play of 'here' and 'there' and of the child's coping with the absence of the mother. By analogy, spatial practices re-enact this play of 'presence and absence' and 'here and there'. The link between spatial practices and childhood experiences is (the) unconscious. So, the child's desire to control the mother plays out as a desire to control space: that is, to render space visible, transparent and readable. The repetition of this inaugural psychodynamic helps explain why, according to de Certeau, the medieval city was imagined and imaged from the point of view of God, from the point of view of a totalising eye, because this perspective spoke of an unconscious wish to 'possess' space. These unconscious interests and desires insinuate themselves into narration and walking: that is, into the production of space.

There is a psychodynamics of place, in this analysis, which involves not only desire, love, loss, the unconscious and megalomania but also movement, seeing, meaning (dreams, language, narration) and bodies. Nevertheless, the production of space is reduced to one psychosexual formant: that is, de Certeau fixes the psychodynamics of place into the Fort/Da game and, thereby, associates space with the mother's body (using a somewhat spurious citation of Freud, 1926b, page 240).[25] Moreover, repression and resistance – as they bleed into the practised place of the city (and the body) – are cauterised: the flow of ambivalence is stemmed by the decapitation of above from below, of Real from metaphorical, of blind love from the mother's body. De Certeau's analysis highlights certain aspects of the psychodynamics of place, but cannot see 'the view from above' as a site of erotics and of resistance (see the discussion of Morris, 1992, above), nor 'the view from down below' as a site of voyeurism and of power. There are, in response, other possibilities for thinking through the psychodynamics of place, if other figures are considered.

In his analysis, strategic elements of de Certeau's celebration of walking in the city appear to be about a particular practice of walking, flâneurie, associated with a particular figure: the flâneur.[26] Thus, de Certeau does not have in mind other kinds of walking: for example, as a practice of power – such as soldiers marching or police patrolling – or as a fixed set of routines – such as going to work, picking up the kids or shopping. De Certeau, then, does not discuss different social and spatial practices of walking, involving who is doing the walking, how and why they are walking, under what

25 Here de Certeau has (unintentionally) smuggled in the familiar co-ordinates of a misogynist, masculinist, heterosexist and phallocentric understanding (see Part II above).

26 This would explain why the title to subsection 2 of his Chapter IX, page 97, is 'The Chorus of *Idle* Footsteps' (my emphasis) – the flâneur was caricatured as an idler – and why the walker is differentiated from the passer-by. Because passers-by are going from one place to another, they can be mapped; whereas walking is another kind of activity, which cannot be represented by a line on a map. Walking – for de Certeau – is both a creative act of writing and a relationship of proximity and distance, just like flâneurie. Indeed, de Certeau describes walking as a long poem which manipulates the organisation of space (page 101). There is a difference between 'walking' and 'flâneurie', however: de Certeau sees walking as a practice of the masses, whereas flâneurie was a definitively élite activity.

circumstances. These issues, for me, suggest a different understanding of the psychodynamics of place. Nevertheless de Certeau's assumption of flâneurie, as his implicit model for the long poetry of walking, is an apt place to start looking for other aspects of the psychodynamics of place, which are actualised through interrelated regimes of vision, power and sexuality. It might be said, following de Certeau's analogy, that lovers only see what they want to see, but they rarely meet on equal terms, and they never occupy the same space.

Flâneurie was (or is) a particular style of walking: the flâneur strolled purposelessly around, looking at things. 'The Flâneur' has become an important 'location' in critical theory because he stands at the intersection not only of class, gender and race relations, and also of art, mass production and commodification, but also of the masses, the city, and the experience of modernity. In his study of Charles Baudelaire (1973), Walter Benjamin suggested that 'the flâneur' was emblematic of urban experience, modernity and the relationship between the intellectual and the world (of things, of the masses). This study locates 'the flâneur' within a specific social formation, time and place: industrial capitalism in nineteenth-century Paris (see Buck-Morss, 1989). As the figure of the flâneur became a leitmotif for the production of art in an age of mass consumption, it has attracted no little critical attention (see Tester, 1994a),[27] but feminists have pointed out that the flâneur was characteristically male and that, therefore, flâneurie did not just tell of urban modernity, but also of the intersection between male power, masculinity and voyeurism (see Wolff 1985; Schor, 1992; E. Wilson, 1992). It is for these reasons that I would like to discuss 'the flâneur', as the embodiment of power-infused spatial practices of walking and looking. Moreover, I have found that discussions of flâneurie commonly deploy the figure of 'the prostitute'. It seems that experiences of masculinity, the city and the capitalist modern throw out relations of sexuality and gender onto the streets. For this discussion, I will be drawing on two resources – Griselda Pollock's (1988b) chapter on the production of feminine spaces and Judith Walkowitz's book on narratives of danger and delight (1992). In the first instance, though, it is useful to quote Tester's neatly distilled description of the hero of modernity – the flâneur:

> The *flâneur* is the man of the public who knows himself to be of the public. The *flâneur* is the individual sovereign of the order of things

---

27 By stressing different aspects of flâneurie, authors provide different characteristics, times and places of the flâneur. He (and, later, she) has been found to be an idle stroller who took turtles for walks in the early nineteenth century; an indoor talker, and prototype of the chattering classes, in the late nineteenth century (and it is at this time that there is a flâneuse – the name for a kind of reclining chair!); a female shopper in a department store around the turn of the century; the contemporary shopper in the shopping mall; and a surfer or lurker in cyberspace (see Tester, 1994a). I am not concerned with these other spatial practices of flâneurie in this chapter, except where these are emblematic of different aspects of the psychodynamics of place.

> who, as a poet or as the artist, is able to transform faces and things so that for him they have only that meaning which he attributes to them. He therefore treats the objects of the city with a somewhat detached attitude (an attitude which is only a short step away from isolation and alienation . . .). The *flâneur* is the secret spectator of the spectacle of the spaces and places of the city. Consequently, *flâneurie* can, after Baudelaire, be understood as the activity of the sovereign spectator going about the city in order to find the things which will occupy his gaze and thus complete his otherwise incomplete identity; satisfy his otherwise dissatisfied existence; replace the sense of bereavement with a sense of life.
>
> (1994b: 6–7)[28]

The flâneur is an ambivalent figure: on the one hand, captivated by the movement and excitement of the urban modern; on the other hand, terrified of being swallowed up by the masses. (An ambivalence which strikingly parallels that of the soldier males discussed in Chapter 6; indeed, it would be possible to read the flâneur as representing a similar psychodynamics of body-ego-space.) Despite the flâneur's constitutive insecurity and despite not placing himself amongst the powerful, this man still occupies sites of authority. These sites are simultaneously close and far – a play of proximity and distance. This man walks the streets in search of new faces, new experiences, yet constantly polices his closeness to those faces, those experiences, never letting himself become embroiled in the daily lives of other people. The spectator marks himself out from the spectacle, never becoming a spectacle himself: he is in the streets, but not of the streets; he is in the crowd, but never of the crowd. In his ambivalent world, the poet male rules the crowd; nevertheless, the long poem is always incomplete and the flâneur is perpetually compelled to immerse himself once more in the masses.

The poet male takes onto himself the right to give significance to the world: he looks at certain things, while other things are deemed not worth looking at; through the eyes, the poet makes things appear and disappear; what he sees, he writes about – and, through poetry, he imposes an order on things (hence, Benjamin's famous remark that the flâneur botonises the asphalt). In this way, the flâneur controls the scene 'because he can and does look just like anyone else, nowhere is forbidden to him; spatially, morally and culturally the public hold no mysteries for the man who is proud of the mystery of himself' (Tester, 1994b: 4). The flâneur seems to be just like everyone else, but he is a spy, a tourist, a detective, a journalist, scrutinising the otherwise alien streets, reporting back on its excessive, exotic, erotic lives. This person, also, is unmistakably marked by his position within gender, class and race relations. Nevertheless, he is not an all-powerful, all-seeing, all-

28 See also Pollock, 1988b: 67.

encompassing male bourgeois white: his location is more marginal; but while he may be in the margins, he is not of the margins. Indeed, it is this poet male's location inside *and* outside power relations that gives him access to the streets, to the crowds, to the erotic underground of city life.

While de Certeau feels that leaving the streets takes you away from a grasp of the hustle and bustle of the masses, the flâneur absents himself through his eyes. This abstraction of experience through the eyes was seen as formative of modern capitalist life by Lefebvre, yet this would now seem to be a very peculiar act: a nineteenth-century Parisian poet male performing a not-quite dominant (perhaps even counter-hegemonic) masculinity. Nevertheless, if flâneurie is a performance – designed to install identity, where there is none; to compulsively repeat the experience of alienation, distance and loss; to give satisfaction, where none can be achieved; and, to forestall death – then it is possible to read this spatial practice psychoanalytically. More than this, it is possible to argue that the flâneur is a masquerade, which acts out its constitutive ambivalence to others, through a play of absences and presences, in the site of others.

Like the Fort/Da game, there is a vulnerability and a megalomania in this situation (as for the soldier males). Moreover, the poet male fantasises himself inside-out, reading his feelings off the spectacle of the crowd. The streets, people's faces and things become blank screens onto which the flâneur's excitements, pleasures and fears are projected. However useful it is to start reeling off psychoanalytic (Lacanian, Freudian, Kleinian) interpretations of the flâneur's supposed neuroses, of his object relations, or of his position within a male gaze,[29] I believe that the scenes of performances are also constitutive of these activities: that is, the spatial practices of voyeurism and megalomania are *also* established through the places they inhabit, through the social relationships which bring them about. While it matters who is walking the streets and who is doing the looking, and why, it also matters which streets are being walked, and how the spatial regime of the visual is constituted.[30]

For Pollock, the streets of nineteenth-century Paris are a site for the remaking of both masculinity and modernity (1988b, page 51). Baron Haussmann was carrying out a new urbanism on a grand scale. Slums were torn down and in their place a modern phantasmagoria was built; its uniform façades of the tree-lined boulevards acted as monuments to the myth of progress; the streets imposed an order, where previously there had been fragmentation; and the 'strategic beautification' of the city bore the moral imprint of the political order (see Buck-Morss, 1989: 88–92). There was a more prosaic sense of order too: the streets formed straight lines between military barracks and working-class areas, and the boulevards were too wide for hastily-erected barricades. Paris had disappeared and reappeared. The

29 For a discussion of 'the male gaze' by a geographer, see Gillian Rose, 1993: Chapter 5.
30 See also Pile and Thrift, 1995b: 39–48.

appearance of the city had changed in other ways: consumption, recreation and pleasure had been mapped into the city streets. The boulevards had become sites of leisure and of the conspicuous spectacle of money (if only in terms of the French state's financial commitment). It is in these spaces that the flâneur flourished.

According to Buck-Morss (1986: 116), the flâneur was the first to be seduced by the world of appearances, by the ever-increasing frenzy of movement of capital, people and commodities, and the commodification of the world. As trapped in the spectacle of the urban modern as the child in the mirror, the flâneur could not see beyond the glass to the bourgeois values etched into the tain of the city. The streets had become a hall of mirrors, of shop windows, which framed, reproduced and distorted every desire and pleasure: this is commodity fetishism, the erotics of consumption, produced *in situ*. The streets had become the interiors of a bourgeois world, where shop windows are mirrors reflecting their desire back, but hiding the soulless heart of commodification, of mass production: the nightmare reality of capitalist dream production (from Buck-Morss, 1986: 111, 122, 129 and 137).

Not everyone had equal access to the streets, nor did everyone share the same license to look; not every greeting in the streets made alienation melt away, nor did every look spell detachment. The flâneur's (like the social reformers')[31] authorisation to loiter in the streets, enjoy looking at the behaviour of others and report on them (as art) both articulates and maintains a particular masculine sexuality, within a Modern sexual economy, within a moral discourse and within a regime of state legislation (see Walkowitz, 1980, 1992).[32] Women did not enjoy this privilege; nor did the working classes. While the flâneur seemingly patrolled this border zone marked out by the production of sexuality, a sexual economy and class relationships, the prostitute actively appeared to cross them. Given the flâneur's seduction by the streets, by commodities, by appearance, by depravity, by vice, it is not surprising that the figure of the prostitute should appear as the emblem of the other side of relations of power, desire and looking. In Benjamin's unpublished Arcades Project, for example, the figure of the prostitute marked the real nature of capitalist labour relations and commodification.[33]

---

31 Walkowitz argues that 'urban investigators not only distanced themselves from their objects of study; they also felt compelled to possess a comprehensive knowledge of the Other, even to the point of cultural immersion, social masquerade, and intrapsychic incorporation' (1992: 20).

32 There is remarkably little work on prostitution by geographers; the two exceptions I know of are Ogborn, 1992 and Law, forthcoming: see also Driver, 1988; Jackson, 1989: 115–120 and Swanson, 1995.

33 Lefebvre makes a more general point about the interrelationship between abstract social relations and the female body:

> In abstract space, and wherever its influence is felt, the demise of the body has a dual character, for it is at once symbolic and concrete: concrete, as a result of the aggression to which the body is subject; symbolic, on account of the fragmentation of the body's

> The prostitute is the ur-form of the wage labourer, selling herself in order to survive. Prostitution is indeed an objective emblem of capitalism, a hieroglyph of the true nature of social reality . . . The image of the whore reveals this secret like a rebus. Whereas every trace of the wage labourer who produced the commodity is extinguished when it is torn out of context by its exhibition on display, in the prostitute, both moments are visible. As a dialectical image, she 'synthesises' the form of the commodity and its content: she is 'commodity and seller in one'.
> (Buck-Morss, 1989: 184–185)

For this argument to make sense, prostitution 'on the streets' of the Modern city has to be different from earlier practices of the so-called 'oldest profession'. The image of the prostitute, it could be countered, had always embodied 'commodity and seller in one' and, although this may make her the archetypal form of the wage labour relation, there is nothing specifically bourgeois or capitalist about this – the same might be said, for example, of soldiering. The link that is being made here is between the commodity, the desire for the commodity, and the desire for sex: and it is into this heterosexual economy of desire, power and exhibition which the flâneur fits. The double-sided relation between the flâneur and the prostitute, then, is about the intersecting relationships of class and gender, of power and desire.

Just as 'the flâneur', 'the prostitute' was constructed as an object of desire and disgust: Cities of Dreadful Delight are at the same time Cities of Dreadful Night. The ambivalent and ambiguous figures of 'the flâneur' and 'the prostitute' are simultaneously the dream and nightmare of bourgeois social reality: the image of a desire which cannot be satisfied, the appearance and commodification of sex, desire and (romantic) love – the poet sells his art, the streetwalker sells her sex (while the worker sells his labour). Nevertheless, this interpretation has overlooked one vital, necessary and irreducible relation: space. Streetwalking is constitutive of both 'the flâneur' and 'the prostitute': it matters that they appeared on the streets, on those streets, for practices of flâneurie and prostitution are simultaneously historically, socially and geographically embedded. Thus, according to Walkowitz, in London,

> the prostitute was the quintessential female figure of the urban scene . . . For men as well as women, the prostitute was the central spectacle in a set of urban encounters and fantasies. Repudiated and desired, degraded and threatening, the prostitute attracted the attention of a range of urban male explorers from the 1840s to the 1880s.
> (Walkowitz 1992: 21)

> living unity. This is especially true of the female body, as transformed into exchange value, into a sign of the commodity and indeed into a commodity *per se*.
> (Lefebvre 1974: 310)

Ironically, both Benjamin and Lefebvre repeat the abstraction and commodification of the female body.

(Ambivalent) Male urban explorers wandered the streets at night engaging themselves in conversations with prostitutes. In tales of these nocturnal activities, the prostitute appeared in two basic guises: first, she was a 'soiled dove' sauntering down the streets along with other prostitutes; second, she was a 'lone streetwalker, a solitary figure in the urban landscape, outside hearth and home, emblematic of urban alienation and the dehumanization of the cash nexus' (1992: 22). Still, prostitutes were 'female grotesques, evocative of the chaos and illicit secrets of the labyrinthine city' (page 22). These women were socially marginal, but they were highly visible: not just in moral and political discourse, but geographically, in the streets, whether they were elegantly dressed walking around fashionable shopping districts, or committing 'acts of indecency' in back-alleyways and in the courts of slums. Prostitutes were caught in the space between classes, between sexes and between looks and, for this transgression, they were subject to intrusive and oppressive state legislation (see Walkowitz, 1980; Ogborn, 1992).

The story, so far, appears to be quite straightforward: the ambivalent flâneur walks the streets creating art from the objects that he encounters; the prostitute is the ambivalent object of the looks of mainly male urban reformers; while the streets are the scene of these encounters. This story would appear to be limited in two directions: first, because it is set in the urban Modern of the nineteenth century; and, second, aside from ambivalence, it lacks an obvious 'psychodynamics of place'. Moreover, for Benjamin, these figures were only two amongst many others which condensed bourgeois social reality:

> The gambler and the flâneur in the Arcades project personify the empty time of modernity; the whore is an image of the commodity form; decorative mirrors and bourgeois interiors are emblematic of bourgeois subjectivism; dust and wax figures are signs of history's motionlessness; mechanical dolls are emblematic of workers' existence under industrialism; the store cashier is perceived as 'a living image, as allegory of the cashbox'.
>
> (Buck-Morss, 1989: 228)

From a different direction, the streets are emblematic of the production of a psychodynamics of place which involves the intersection of desire, voyeurism and commodification, of race, class and gender. The streets are part and parcel of an urban imaginary, populated by figures who are not simply metaphorical or allegorical, but also flesh and blood. If the gambler and the flâneur are emblematic of empty time, then they also signify the production of particular sites – the gambling den, the shop window, the street; if the prostitute is emblematic of the commodity, then she is more than a commodity – she is much more than just sex for sale (see Welldon, 1988; Ogborn, 1992; Law, forthcoming). If the streets are scenes of desire and disgust, then they are also constitutive of those desires and disgusts – they are a hall of mirrors, a shop window, the visible excess which the bourgeois home was designed to hide away.

Pollock and Walkowitz argue that, in the nineteenth-century city, the intersection of bourgeois and masculine values and power produced a division between public and private space: men occupied and controlled the public sphere, while women were confined to the private sphere – the flâneur emphasises this social and spatial division, while the prostitute transgresses it. Social values and social hierarchies were not just mapped across the city, the city opened up people to inspection. Men and women in Paris and London were mapped according to their location: who you were was read off from where you were, who you appeared to be and where you were seen – and the flâneur and the prostitute are emblematic of this double/d meaning of location as both place and social standing – but the social and spatial script was neither well written nor easily read. Thus, masculine sites of power were privatised, while other women also walked the streets when, for example, they went to work. While the flâneur may in fantasy have controlled the scene, life continued without him. The surveillance and regulation of urban socio-spatial practices could not contain the spaces of the subject.

It is commonly argued that the flâneur marks the spot of a powerful, active, voyeuristic male gaze (and, by analogy, represents all men); while the prostitute is the passive object of desire who only gains meaning through displaying herself and being bought (and by analogy is the fate of all women). Even so, they are also markers of the difficulties of locating people in their proper place: each moves between locations, turning up where they 'should not be'. Indeed, they matter precisely because they are both 'out of place'. Moreover, each is both active and passive; each is scarred by the vicissitudes of desire, disgust and power. In this sense, it might be argued that they are the 'contents' of a nightmare that proves the urban has an unconscious, an underground erotic life. Furthermore, the flâneur was himself a marginal figure, while the prostitute was never merely a submissive object of desire and disgust: prostitutes may not have walked the streets on the same terms as men, but this did not necessarily make them passive victims (see E. Wilson, 1992; Ogborn, 1992). Each was involved in the grotesque psychodrama of Modern life, each involved in the production of the look of the other, played out in the sights produced by urbanism: the street. Neither figure is fully private nor fully public, their identities are played out *in situ*: the social masquerade of the streets. And ambivalence pervades city life – an unfilled promise, a death threat. So, more than once, there is a psychodynamics of place.

CONCLUSION: EROTIC LIVES IN THE CITY

Sally Munt says:

> Brighton introduced me to the dyke stare, it gave me permission *to* stare. It made me feel I was worth staring at, and I learned to dress for the occasion. Brighton constructed my lesbian identity, one that was given to me by the glance of others, exchanged by the looks I gave

them, passing – or not passing – in the street.

(Munt, 1994: 62)

Munt's experience of Nottingham was a complete contrast. She felt herself to be the object of a heterosexual gaze which made her into a victim. She actively sought out the identity of a flâneur, a performance which could empower her and enable her to take control of her desire, her look: 'strolling has never been so easy, as a spatial zone, the lesbian city, opens up to me' (1994: 2). It is not just that certain spaces are eroticised, but that there is an eroticisation of urban lives in the city, and that the erotic lives in the unconscious of the city. Brighton gives permission, while Nottingham prohibits: the street is the site of a masquerade, while social performances are circumscribed by relations of power. There is a double-edginess to urban living – neither fully inside nor outside, private nor public, closed nor open, real nor metaphorical. And Munt's account also highlights the intense rigidity of the heterosexual, gender and class grids through which the figures of the flâneur and the prostitute have been interpreted. She shows that the positions which people occupy are never so fixed; she shows that the sexual politics of the look can involve other pleasures.

In the city, people live their lives through their object relationships: that is, through an urban relationship to the world. This relationship, to take directly from a psychoanalytic source, 'is the entire complex outcome of a particular organisation of personality, of an apprehension of objects that is to some extent or other phantasied, and of certain types of defence' (Laplanche and Pontalis, 1973: 277). It is not so strange, then, to think as Pocock does of 'city personality', although he did not make anything of the ambiguity of this phrase. Through fantasy, whether conscious or unconscious, the urbanised subject creates an imaginary urban landscape, which is constructed partly by the material of the city, partly by the modalities of identification, partly by defensive processes and partly by the 'contents' of the unconscious.[34] It is through these simultaneously real, imagined and symbolic spatialities that spatial practices – of the city, of the body – develop. From this perspective, the mental map of the city is about 'imageability' and about some kind of 'embodied knowing'; but, as important, are the subject's relationship both to imagery, symbolism and some kind of 'reality', and to the play of absence and presence, of voyeurism and megalomania, of desire and disgust, of social masquerades, of power and meaning. In this sense, the mental map of the city is not a product, but a process: the psychodynamics of place.

The psychoanalytic cartography of bodies and cities has taken in a variety of sites, but none of these places have been innocent of power relations. In part, this analysis has shown that power is never fully external to the subject.

---

34 Identification also includes incorporation, idealisation, introjection and internalisation; while defence mechanisms include repression, sublimation, and reversal into opposites.

It has to be added that subjectivity is neither simply produced nor determined by its location within multiple, interrelated modalities of power, such that the subject experiences these relations in often surprising, and sometimes perverse, ways. From colonial desire to body-ego-space, from the Manhattan skyline to streetwalking, there must also be a place for the politics of subjectivity.

CONCLUSION

8

CONCLUSION

places for the politics of the subject

THE SAME SPACE CANNOT HAVE TWO DIFFERENT CONTENTS

In 'Civilization and its discontents' (1930), Freud discusses the ways in which the child moves along the path of its development. The child is learning to distinguish between an internal and an external world, through its body and its perceptions. Indeed, the infant is becoming a subject through the projection of sensations onto the surface of the body, through its embodied knowledge of the realities of the world outside. At a certain point, Freud begins to discuss memory-traces in the mind. He is searching for a metaphor for the indestructibility of psychic reality. Freud selects the example of 'the history of the Eternal City' (page 256): Rome. He then traces out the founding of Rome, its domination by the Caesars, and its transformation in the Renaissance. Beneath the soil of the contemporary city can be found the memory-traces of ancient Rome, but this history is also preserved on the surface. The Eternal City is like the Eternal Mind. Temples, walls, piazzas are built and torn down as Caesars rise and fall. The history of the child's experiences is never eradicated, it is to be found in the archaeology of the city of the mind, in the dynamics of the present and in its ruins. Then Freud gives up: 'There is clearly no point in spinning our phantasy any further, for it leads to things that are unimaginable and even absurd. If we want to represent historical sequence in spatial terms we can only do it by juxtaposition in space: the same space cannot have two different contents' (1930: 258).

Freud then begins to wonder about the specific usefulness of choosing a city as the analogy of the mind. He abandons the metaphor on the grounds that the city is constantly changing even during its most peaceful periods: buildings are forever being torn down and replaced, whereas the mind develops in response to anxiety, threat and conflict. Meanwhile, most cities have been damaged by destructive forces, whereas the mind suffers conflict even as the brain remains physically intact. Neither Mind nor Body are exactly like the Eternal City. Perhaps, though, the analogy would have held better had Freud thought about the struggles and conflicts within the urban which are traced through its fabric, through the production of its space. Had he been around, Mumford could have told Freud that the city itself develops in relation to anxiety and threat, though this is most true of walled cities (Mumford 1961). Maybe the analogy would have held if Freud had developed further the idea of city sites as representing particular ego-feelings (Temple, Citadel, Forum, and so on). Either way, Freud has not quite finished with city as a metaphor-concept for specific mental processes.

Although Freud could not see any way around it, he thought that culture and civilisation inevitably thwart the individual. There is a profound doubleness to civilisation: on the one hand, it binds people into it through ties of love; yet, on the other hand, it prevents the free expression of the libido. Civilisation depends on love, yet frustrates it. Love and Frustration go hand in hand. So, the programme of building civilisation is perpetually threatened by overspill of frustration into destructiveness and aggression. Freud associates the constitutive double-edged sword of civilisation with the opposed, conflicting tensions within the individual: the sexual versus the destructive drives. These are, famously, mythologised as Eros and Thanatos. Essentially, civilisation must spoil the designs not only of Eros, but also of Thanatos. It is the necessary repression of these instincts that, for Freud, lies at the heart of people's misery and unhappiness. Freud has, however, not projected psychic processes onto the character of civilisation, nor reduced the individual's impulses and feelings to their cultural determinants. In order to specify this particular relationship, Freud uses another analogy. Through 'conscience' and guilt, 'civilisation ... obtains mastery over the individual's dangerous desire for aggression by weakening and disarming it and by setting up an agency within him to watch over it, like a garrison in a conquered city' (1930: 316).

The garrison in the conquered mind is the super-ego. Civilisation works not just because élites oppress the masses, but also because individuals repress themselves. Civilisation does not merely conquer bodies – individuals become subjects through the installation of civilisation within them. The conclusion that so many critical theorists, from the Frankfurt School to feminism, have come to is that freeing the body *also* requires the expulsion of the garrison which controls interior spaces. From this perspective, the question is raised as to how this can be done and whether in fact some kind of authority is necessary to keep unruly elements in the city in their proper place: in sum, is an 'unrepressed' civilisation possible?[1] I am not going to pursue these questions in this way, rather I am going to sketch out the spatialities which are involved in Freud's Roman analogies. By disclosing these spaces it is possible to specify the 'dialectics' between the individual and 'civilisation', and also to show how subjectivity, space and power are intertwined. I will deal with these issues in this section. In the two closing sections, I will use this discussion to inform the politics of placing the subject:

1 These questions have been most fully explored by the Frankfurt School, see Held, 1980, and by feminists, see Mitchell, 1974.

initially, through a summary of the political aesthetics of Fredric Jameson and, finally, through Frantz Fanon's description of an encounter. Before this, let me return to the spaces of mind, body, city and civilisation.

Although Freud is taken by the idea that psychoanalysis might be an archaeology of the mind, he rejects the idea that the mind is like Rome on spatial grounds – 'the same space cannot have two different contents'.[2] The issue here concerns what kind of spatial imagination Freud is using to conceive the mind–body. Freud seems to be saying that the experiences of the child form an archaeology of the mind where these exist both in different and in the same spaces. Different parts of the mind are interwoven, yet fragmented and distinct. Importantly, Freud is rejecting a geological model of space-time where experiences are like sediments which are overlain by subsequent events.[3] Instead, he is insisting that the space-times of the mind are constantly located, dislocated and relocated in sly, shifting and concrete sites. The topologies of the mind are partitioned by unfolding border lines, but there are constant exchanges across and within the contact zones between them, which constantly deterritorialise and reterritorialise feelings, thoughts and actions. Freud's understanding of the mind–body remains stubbornly spatial, forever walking the tightrope between ideas which are too slow and too fast.

From this perspective, two related questions beg the in/adequacy of the city metaphor-concept: first, can space have two (or more) contents and, second, what are the spaces of the psycho-body? The City, to push it as a metaphor for power and civilisation as well as mind and space, can be made to show that space can have two or more contents, where the specific relationships within and between sites are uncovered. Buildings, for example, mobilise, organise and house different times, spaces and functions. Meanwhile, the idea of 'the underground' can be taken to mean many different kinds of things. And conflict, compromise and compassion are sharply juxtaposed in the metropolis. The spaces of the urban are analogous to the spaces of the mind: conscious, preconscious, unconscious; with shifting, positioning and fighting between them in a struggle for control and expression. Such musings, however, seem vague and unsatisfactory. There are other approaches to this problem of the relationships between the archaeology of space, the spatialities of the mind–body and the spaces of power.

If Freud's rejection of Rome as a model of the space of the mind marks a rejection of a static, passive and undialectical notion of space (which is encapsulated and exemplified in his notion of the unconscious as a space without time) and if his use of the idea of a garrison in a conquered city speaks of the ways in which power installs subjectivity through the surveillance of interior spaces, then there are here specific ways of describing

---

2 For a related reading, see Kirby, 1996, Chapter 3.

3 I have called this a 'geological' sense of space-time, but it could be equally well described as Newtonian, or Cartesian, or Darwinian.

the power-laden 'dialectics' which graduate the subject, the spatial and the social. More than this, *The Body and the City* has shown that these 'dialectics' can be broken down into the following components: identification, incorporation, idealisation, introjection and internalisation – and their associates projection and abjection. Each term is both a specific 'dialectic' and a specific space.

The child becomes a subject in relation to the body and the city through its use of, for example, incorporation and abjection to 'manage' its feelings, thoughts and behaviour. While a garrison may be set up to police the child's adoption of strategies of identification and disidentification, the infant will also be able to select and choose alternative tactics for becoming a subject, within the circumstances it finds itself. Objects and ideals may be presented to the child, but the playing out of the interaction between fantasy, body and space will be for the child: 'you can lead a horse to water, but you can't make it drink' was not an expression designed to describe the behaviour of horses. The child's tactics produce many spaces in the same space: spaces projected from the inside onto the surface of the body; garrisons which police an occupied city full of wants and tantrums; objects forever preserved within the body–mind; outsides inside; insides outside; border zones; lines of demarcation; ideals; and so on. None of these spaces is innocent of others. None of these spaces is innocent of power and resistance: resistance (for the moment) in a double sense of 'keeping away from authority' and of 'lying in wait to ambush authority'.

There is constant struggle through these spaces to play off proximity against distance, as exemplified in the Fort/Da game. Part III provided evidence of the ways in which spaces are, for example, incorporated into the body–mind, such that attacks on the nation are experienced as violations of the body, while female bodies of others are abjected; a complex where there is a preoccupying desire to expel a disgusting presence within the body, which because it is internal cannot be eliminated except through fantasy, by the elimination of the presence of others. The purification of space becomes an obsession where abjection takes hold: the city is purified as the nation is purified – the history of the present testifies to the sickening ferocity with which these feelings can be prosecuted – on bodies, in cities. These are all aspects of the psychodynamics of place.

In this book, I have suggested that a 'psychoanalysis of space' might also draw on a Freudian re-reading of Lacanian psychoanalysis. Six 'psychogeographical' issues were identified. These concerned:

1 the object relationships between the body, the 'me'/'I' and the subject's place in the world;
2 the sense that subjectivity is played out through repression and resistance;
3 the ways subjects, objects and spaces are constituted out of their partiality, their duplicity, their virtuality and their supposed truth;

4 an awareness that subjects act out their subjectivity as a situated and repeated performance;
5 a differentiation between dissimilar (psychic, social) spatialities and the often conflicting relationships within and between them; and
6 an alertness to social sanction, social power and the possibilities of radical politics.

I have used these ideas in Part III to think through the relationship between meaning, identity and power: a number of co-ordinates have been traced, through spatial points of reference such as 'constellations', 'mirrors', and 'masquerades'. I have traced these relationships through two sites, the body and the city. These discussions have not sought to be exhaustive nor have they been an attempt to paste a master-narrative over the surface of space. Moreover, the geographies that have been mapped have avoided the reduction of one term to any other term, whether this term is the phallus or the dialectic. The overall impression that these geographies give is of the dynamic interactions between mutually recombining modalities of subjectivity, spatiality and power.

Whether it is the 'body politic' or on 'the streets', I have suggested that the subject is placed by, places themself in relation to, and alters simultaneously Real, Imaginary and Symbolic spatialities. To be an urban subject means something – and the geography of the city means something: whether it is in Freud's uncanny story of dreadful delight or in the freedom it gives for people to lose themselves and look anew. This sense of space gives Park's suggestion that the city is 'a state of mind' an added resonance.[4] In different circumstances, Minnie Bruce Pratt has described the way the city becomes a state of mind this way:

> I was shaped by my relation to those buildings and to the people in the buildings, by ideas of who should be working in the Board of Education, of who should be in the bank handling money, of who should have the guns and the keys to the jail, of who should be *in* the jail; and I was shaped by what I didn't see, or didn't notice, on those streets.[5]
>
> (Bruce Pratt, 1984: 17)

The streets become a map of visible and invisible relations of meaning, identity and power into which the subject is placed and has to find their way around – and possibly, one day, to escape. From this perspective, space has

4 See Park, 1925: 1. Following this line of argument, Park adds that the city is not 'merely a physical mechanism and an artificial construction. It is involved in the vital processes of the people who compose it; it is a product of nature, and particularly human nature' (page 1). I think that much could be learnt from a reassessment of the Chicago School's early work, though see Sibley, 1995b.
5 For a fuller reading of this article, see G. Rose, 1993: 156–158.

two and more contents, in part, because it contains the psychodrama of everyday life.

At the outset of this book, I began with geographers' desire to explore the *terrae incognitae* of 'the hearts and minds of men'. Some thought that this desire might be grounded in a geographer's libido, while for others 'geography' was founded by the universal human need to know their way around their world, otherwise people would be unable to control basic resources, such as food, warmth and shelter. For me, the justification for placing the subject in 'psycho-geographical' maps of the world is to enable a progressive politics. Currently and commonly, spatial metaphors have infused discussions of radical politics (see Pratt, 1992; Smith and Katz, 1993). To conclude this book, I would like to look to two places where psychoanalysis has informed a discussion of subjectivity, politics and space. The first is Jameson's suggestion that what is needed to inform radical politics is an aesthetics of cognitive mapping (1991). The second is Frantz Fanon's analysis of the colonisation of interior spaces (1952).

PLACES IN COGNITIVE MAPS

The idea of cognitive mapping is hardly marginal in Jameson's lengthy dissection of postmodernism. It is used to conclude not only many chapters within the book, but also the book as a whole. The argument of the book depends on a renewed sense of the importance of space, derived mainly from Louis Althusser, Henri Lefebvre, Kevin Lynch and Ed Soja. The provision for an aesthetics of cognitive mapping is a political moment: it is this which will enable the individual subject to intervene in the world, to make it a better (if not perfect) place (see Chapters 2 and 7 above). This sense of political practice arises because Jameson detects a shift in the logic of social organisation which is manifest in daily life, psychic reality, space and social structures: 'I think that it is at least empirically arguable that our daily life, our psychic experience, our cultural languages, are today dominated by categories of space rather than categories of time, as in the preceding period of high modernism' (Jameson, 1991: 16).

The logic of social organisation has mutated into a 'postmodern hyperspace' which exists at a level which is impossible for the individual to understand. More than this, the individual subject is totally alienated from their surrounds. People are unable to place their bodies, their selves, in relation to the real determinations of social life. This 'postmodern hyperspace' subsists at a global scale and it is impossible for people to hold this world in their minds: the body, the self, the system cannot be cognitively mapped and, so, people cannot find their place in 'a mappable external world' (page 44). For Jameson, there is an alarming disjuncture, or gap, 'between the body and its built environment' (page 44). This gap, or alienation, between the external world and the ability to hold that world in the mind reminds Jameson of Kevin Lynch's study of *The Image of the City* (1960) (pages 51–52 and 415–416): 'The alienated city is above all a space in which people are unable to map (in their minds) either their own positions or the urban totality in which they find themselves' (page 51).

Like a cityscape that has become illegible, without clear markings for where the individual is, so the contemporary has lost its landmarks and boundaries. In a sense, Jameson's postmodern hyperspace is a kind of urban space without a geography.[6] This geography-less space submerges the subject in a space without co-ordinates, in a space which abolishes any idea of distance. Without some sense of getting a distance on things, any possibility of critical thinking is permanently immobilised. With no geography, no space, no confines, no prison, there is no way out – unless it is possible to map the prison and chart escape routes. It is this hope that leads Jameson to posit a reinvigorated aesthetics of cognitive mapping. The lesson has been learnt from Lynch:

> Disalienation in the traditional city, then, involves the practical reconquest of a sense of place and construction or reconstruction of an articulated ensemble which can be retained in memory and which the individual subject can map and remap along the moments of mobile, alternative trajectories.
>
> (Jameson, 1991: 51)

The cognitive map is not a replica of the external world, it is a means of taking control of the world and making the world anew. Its location in a radical project is further specified in relation to the work of Althusser.[7] Jameson still goes along with Althusser's definition of ideology as 'the representation of the subject's *Imaginary* relationship to his or her *Real* conditions of existence' (Jameson, 1991: 51).[8] Jameson plans to revitalise Althusser's formulation in two ways: first, by linking it to the notion of the cognitive map; and, second, by recovering its implicit use of Lacan's Real, Imaginary and Symbolic orders. Following Lynch and Althusser, Jameson suggests that the cognitive map should enable individuals to represent their situations in relation to their real circumstances within the ensemble of social structures. In this sense, cognitive mapping is designed to uncover both the real determinations of global social relations and the individual's place and position within them. It requires 'the coordination of existential data (the

---

6 It wouldn't be too far-fetched to suggest that 'postmodern hyperspace' could be renamed 'The Bonaventure Hotel' or 'Los Angeles'. Much has been written about the quintessential postmodern space of the Bonaventure Hotel, but I was very disappointed to find that its architecture is no more complicated or disorientating or postmodern than that of 'premodern' Warwick Castle. And Los Angeles has buses. Don't believe the hype(r).

7 It is interesting that 'space' and 'Althusser' seem to be constantly juxtaposed in Jameson's work. The other co-ordinates of this map are, in terms of this book at least, familiar: Lefebvre, Lynch and Lacan (see also Gregory, 1994: 140, notes 149 and 150, and elsewhere).

8 My translation of Althusser reads: 'Thesis 1. Ideology represents the imaginary relationship of individuals to their real conditions of existence.' I presume Jameson's capitals and italics are designed to emphasise the links between Althusser's formula and Lacan's orders. I suspect that the lack of dynamism in Jameson's model stems from his underlying sense of space as being a passive backdrop to social relationships (see Keith and Pile, 1993a: 4).

empirical position of the subject) with unlived, abstract conceptions of the geographic totality' (Jameson, 1991: 53).

This formulation relies on opposing abstract space and personal space, but Jameson disrupts this Althusserian binary schema by introducing a 'symbolic' space. In Althusser's scheme experience and ideology are inextricably intertwined, while the real determinants of everyday life happen behind the back of subjects and are only recoverable through science. Ideology versus Science. While Althusser's formula is useful, maintains Jameson, for mapping the place of the individual in relation to the class realities of abstract space (local, regional, national and so on), the ideology/science dichotomy not only separates experience from abstract social relations, but also reduces it to ideology. Following Lacan, Jameson argues that there is a void at the centre of the concrete subject. This void off-centres subjectivity and prevents the subject from being totally immersed in ideology. In this way, Jameson breaks the link between experience and ideology. Now Althusser and Lacan can be brought into harmony: Science = Real; Ideology = Imaginary; Experience = Symbolic (pages 53–54).[9] This triangulation, for Jameson, creates the possibility of new cartographies of subjectivity (the Imaginary), the ensemble of global social relations (the Real) and the ways in which the gap between them is papered over by conscious and unconscious representations (the Symbolic).[10]

As Jameson stresses, this reconceptualisation is political: it is an aesthetic which subverts dominant representations of space (an argument which connects to Lefebvre's three-fold schema). Cognitive mapping intervenes in the process of bringing individuals into subjectivity by changing the Symbolic, by enabling new cartographies of the global space. For Jameson, new forms of spatial representation will enable people to build a map of abstract or multinational or corporate space or postmodern hyperspace, to get it 'in our mind's eye' (page 127). Once a mental image of this alienated and alienating space has been constructed, Jameson believes, it will be possible to reconquer space, to reclaim memory traces and to set off on alternative trajectories. Following Lefebvre, Jameson argues that what is needed, in order to help find sites of resistance, is 'a new kind of spatial imagination capable of confronting the past in a new way and reading its less tangible secrets off the template of its spatial structures – body, cosmos, city, as all those marked the more intangible organization of cultural and libidinal economies and linguistic forms' (Jameson, 1991: 364–365).

If space is the sign of our times and if modalities of power are constitutive of,

---

9 For an alternative permutation, see Zizek, 1989.

10 See Jameson, 1991: 415–416. Problems with this kind of one to one conceptual remapping were encountered in the Conclusion to Part II above. Intriguingly Jameson's formula is almost behavioural geography. The difference is that 'ideology', rather than 'the image' or 'attitude', spans or maps the difference between the perceptual and the phenomenal environments. For me, this 'coincidence' is founded on their shared assumption that space is a passive backdrop to the production of subjectivity.

and constituted by, space, then it seems reasonable to proclaim the need to 'make meaningful space'. It is possible, from my perspective, to take this claim a little further. Jameson's sense of space is profoundly undialectical, passive and fixed. His argument that power and abstract social relations are evaporating into ever larger spatial scales increasingly renders an understanding of place and of experience useless. As place is dissolved, political action can only occur on the global stage, at the level of aesthetics. For me, there are other possibilities: the cognitive map may be conceptualised in other ways.

Later in his discussion, Jameson takes on board Deleuze and Guattari's sense that people occupy multiple subject-positions and leads this into a discussion of Althusser's idea of 'interpellation,' whereby the subject comes to occupy the place that others see them in. Through being addressed, the subject is marked as belonging to a particular place in society. The subject is provided with an image of the group to which they belong and through this image they become a subject. The moment is profoundly dialectical, in the sense that Lacan's mirror phase is dialectical. Child sees image, image sees child; child becomes image, image becomes child. In this formulation, the subject is nowhere because other spaces are not involved. As I have shown in Part II, however, spatiality is constitutive of these relationships, while the modalities of identification further upset any sense that space is an undialectical, unmediated reflective surface.

Jameson uses the idea of cognitive mapping to suggest new cartographies of global and abstract social relations; from my point of view, cognitive mapping should provide new maps of the coexisting Real, Imaginary and Symbolic spatialities which are constitutive of, and constituted by, subjectivity, space and power. As political practice, it would be predicated on an understanding of the interactions between these spatialities, both within and between individuals and groups, and it would not confine itself to the task of placing the individual in global chains. In this way, it would be possible to think through people's multiple subject-positions, but without placing these positions on a static, undialectical and fixed map of abstract space.

This reminds me of de Certeau's sense of spatial practice as a political intervention (de Certeau, 1984: 97–102). He suggests that 'walking' remaps space in ways which render space visible, create new paths through space and are enduringly reversible or flexible. By finding alternative routes, de Certeau intends to subvert the institution of spatial practices as 'citation' of the already there. For me, this indicates that cognitive mapping must work beyond the spaces of 'the template', beyond the grave of the Xerox machine of meaning, identity and power. If this argument holds, then cognitive mapping is not about putting clearly identifiable bodies or precisely defined power relations into their proper place, but instead about a constant struggle to find a place, a place which is not marked by the longitude and latitude of power/knowledge.

Furthering Jameson's sense of cognitive mapping can take the idea of the subject, politics and space a long way, but it is also possible to address this question from another point of view.

YOU DO NOT SEE ME FROM THE PLACE I SEE YOU

If mental processes are somehow analogous to the city, then its location within an imperial imaginary needs to be foregrounded. Rome was the heart of empire, while the garrison polices the conquered city.[11] The mind–body is in the place of the colonised: invaded, conquered, unable to speak in its own language, constantly policed by an all-seeing eye, suffering the whims of an arbitrary power and threatened by barbaric punishment. Subjectivity is a grotesque psychodrama. The clear danger of this description is that it eliminates the differences between kinds and scales of terror. Not everyone is in the same place. And it is in encounters between people that these differences become sharpest. The work of Frantz Fanon begins to chart the ways in which people (fail to) recognise others as they react to (visible) differences. The political implications of this narrative lead in two associated directions: first, to an understanding of the role of psychodynamics in the 'misrecognition' of visible differences;[12] and, second, to the recognition of the psychodynamics of place in the production of a politics of 'arbitrary closure' or 'strategic essentialism'.[13]

Black Skin, White Masks (1952) is an extraordinary book, though its interpretation and significance is highly contested (see Mercer, 1995, for a recent intervention). Here, I intend to read perhaps the most well-known of Fanon's anecdotes,[14] 'Look, a Negro!', partly because this story marks a sharp change of direction in Fanon's (dialectical) argument. So far, Fanon has been thinking about 'the fact of blackness', but how this 'blackness' is apparent only in relation to 'the white man' (1952: 110). He has been trying to think about the ways in which it is possible for the black man to meet the white man as an equal. Although there are no logical grounds on which it can be sustained that the black man is inferior to the white man, the real world challenges the claims of Reason (page 110). Neither white nor black exist except in relation to one another, yet it is the white man who is the Master and the black man who is other.

Fanon identifies the fact of blackness as being the badge of difference and inferiority in a white-dominated world. Colonial power works through a corporeal schema, which grades bodies according to colour. The colonial situation puts a mirror up to the face of the black man and the reflection tells

---

11 Fanon cites René Ménil as saying slavery involved, 'the replacement of the repressed [African] spirit in the consciousness of the slave by an authority symbol representing the Master, a symbol implanted in the subsoil of the collective group and charged with maintaining order in it as a garrison controls a conquered city' (Fanon, 1952: 143; citing Ménil, 1950). Fanon seems unaware that the 'garrison' metaphor is Freud's, but in any case abandons the idea of an essential African spirit in favour of 'history' and 'situation'.

12 There is a link here to Freud's discussion of the narcissism (and vehemence) of minor differences, see Chapter 4 above.

13 These phrases are Stuart Hall's and Gayatri Chakravorty Spivak's respectively.

14 With Nigel Thrift, I have discussed this story before (1995b). In this interpretation, I pick out different threads (and correct an error of mine).

him that he is inferior and other.[15] The effects of this are far-reaching: it scripts consciousness, installing itself in the unconscious (as Fanon reveals through the interpretation of the dreams and fantasies of his patients): despite his sometimes conscious wish to be (acknowledged as) black, Fanon says that 'out of the blackest part of my soul, across the zebra striping of my mind, surges this desire to be suddenly *white*' (page 63). Throughout *Black Skin, White Masks*, Fanon makes deliberate use of psychoanalysis to argue that the black man, by identifying with – and desiring – the position and power of the white man, ends up by seeing himself as 'not-white', 'not-Master' and 'nowhere'. The 'black' man is alienated from himself in the grotesque corporeal schema of colonial living:[16]

> movements are made not out of habit but out of implicit knowledge. A slow composition of my *self* as a body in the middle of a spatial and temporal world – such seems to be the schema. It does not impose itself on me; it is, rather, a definitive structuring of the self and of the world – definitive because it creates a real dialectic between my body and my world.
>
> (Fanon, 1952: 111)

The black/white grid of meaning, identity and power is not imposed from the outside, but is inscribed in the movements of people, in their actions, thoughts and feelings.[17] This grid not only defines the visibility of the body, and territorialises the body, but it is also woven by the white man 'out of a thousand details, anecdotes, stories' (page 111). Anticipating those who argue that identity is a repeated stylised performance, Fanon slows this entertainment down to the heavy burden of moving through an occupied territory under the constant scrutiny of the Master's blue eyes, which are both fearful and full of fear. The definitive 'performance' of self is placed in the middle of a 'real dialectic' between the (tacit) self, the (seen) body and the interventions of the external (colonial) world:

> 'Look, a Negro!' It was an external stimulus that flicked over me as I passed by. I made a tight smile.
>
> 'Look, a Negro!' It was true. It amused me.

---

15 The analogy of 'the mirror' is not idle; Fanon is consciously drawing on Lacan (page 160, note 28; see also Mercer, 1989). Further, in his essay on Fanon and black art, Mercer (1995: 29) describes a photograph, *Mirror, Mirror* (1987), by Carrie Mae Weems. The image depicts a black woman looking into a mirror. She does not see her reflection in the frame; instead there is another figure, dressed in white and holding a star. The caption reads: 'Looking into the mirror, the black woman asked, "Mirror, mirror on the wall, who's the finest of them all?" The mirror says, "Snow White, you black bitch, and don't you forget it!!!"' The grotesque psychodrama of the corporeal schema which Fanon identifies is not confined to colonial life.

16 To echo the words of Homi Bhabha, 1986: 71.

17 In this Fanon reveals his existentialist predisposition (Sartre and Fanon admired each other's work). Earlier he writes: 'Every one of my acts commits me as a man. Every one of my silences, every one of my cowardices reveals me as a man' (1952: 89).

> 'Look, a Negro!' The circle was drawing a bit tighter. I made no secret of my amusement.
>
> 'Mama, see the Negro! I'm frightened.' Frightened! Frightened! Now they were beginning to be afraid of me. I made up my mind to laugh myself to tears, but laughter became impossible.
>
> (1952: 112)

Fanon has been, in Althusser's/Jameson's understanding, interpellated by the Imaginary corporeal schema of the Real white conditions of existence. He cannot laugh because he knows that the child's fear is based on the 'historicity' of the black body. Fanon is made visible by the skin of his body, but cloaked in the legends and anecdotes that envelope the black body. He is simultaneously visible and invisible, marked and erased, certain and uncertain – he certainly has a black body, but there is a deep uncertainty about what this might be (see Bhabha, 1990: 44). His body has been placed by the white boy into a racialised and racist corporeal matrix.

> In the train it was no longer a question of being aware of my body in the third person but in a triple person. In the train I was given not one but two, three places. I had already stopped being amused. It was not that I was finding febrile coordinates in the world. I existed triply: I occupied space. I moved toward the other . . . and the evanescent other, hostile but not opaque, transparent, not there, disappeared. Nausea . . . .
>
> (1952: 112, ellipses in original)

If Freud has rejected the idea that 'the same space cannot have two different contents', then Fanon has substantiated this suspicion. For Fanon, the black body exists in three spaces: the same space has three different contents, and more. One way to understand this situation is to think of the person: the first person, the person speaking; the second person, the person spoken to; and the third person, the person spoken about. Another way to think about this concerns the body: the physical body, the body marked by the encounter with the other, and the historicity of the body. One space: three persons: three bodies. Fanon had stopped being amused, not because he had lost his place in the world, but because he had found it: he was the person spoken about, the black body, the body which carries the burden of white fears and white hatreds.[18] One space: three contents. Haunted by the presence of white ghosts, Fanon is sick.

Fanon is 'completely dislocated' by this experience. He has been separated from his body; his body has been incarcerated within the prison-house of white desires, fantasies and fears. Fanon is split, both phobia and fetish (see Bhabha, 1986: 78). He is aware of being severed from his body. This amputation makes Fanon endure 'a haemorrhage that splattered my whole

18 Fanon says: 'I discovered my blackness, my ethnic characteristics; and I was battered down by tom-toms, cannibalism, intellectual deficiency, fetishism, racial defects, slave-ships, and above all else, above all: "Sho' good eatin'"' (1952: 112).

body with black blood' (page 112). His body, in three places, is not allowed to be equal to the white man's: across three lines, he suffers the deadly cuts of the racist grid of meaning, identity and power: 'My body was given back to me sprawled out, distorted, recolored, clad in mourning in that white winter day. The Negro is an animal, the Negro is bad, the Negro is mean, the Negro is ugly' (page 113).

Violence, fear, desire, hatred swirl around poles of absolute difference marked by skin: where 'white' is as good as Snow White and 'black' is a place of dark secrets. As Fanon moves around the world, he feels that he is 'being dissected by white eyes, the only real eyes' (page 116). Through naked eyes, Fanon is *fixed* into place, into his body (see Mercer, 1989: 193). There is nothing *essential* about this. There is no Reason for it. As Fanon astutely observes later, 'The Negro is not. Any more than the white man' (page 231). Nevertheless, these opposed, though interdependent, pairings become the hard coordinates of oppression and repression. Desire and fear, identification and abjection go hand in hand.

> 'Look how handsome that Negro is! . . .'
> 'Kiss the handsome Negro's ass, madam!'
> Shame flooded her face. At last I was set free from my rumination. At the same time I accomplished two things: I identified my enemies and I made a scene. A grand slam. Now one would be able to laugh.
> (1952: 114)

Fanon completes the scene with a scene. He shames the madam by putting his 'ass' in her face. The turn has been completed in Fanon's dialectic: he has tried Reason and failed to convince the white world of his equality; next, he will try Unreason – black magic – to outflank the white world, but this ultimately fails too; then, Fanon finds black history, black civilisation, and it is on these grounds that it is possible to show equality with white civilisation. Fanon can trump France by overlaying it with Egypt. Even this twist of the dialectic is unsatisfactory, Fanon wishes to escape history, to free himself from the slavery of Slavery. Hope, for him, lies both in continuing to ask questions and in opening the mind to the presence of others. These are significant ideas, and I will return to them, but first I would like to set the scene.

When I first read Fanon's book, like most I was struck by the ambivalence of identification and desire: identifying with/against, and desiring/hating, within/across boundaries of 'race' and 'sex' (see also Fuss, 1995; Mercer, 1995). An intricate 'cognitive map' of meaning, identity and desire is traced out through a dialectical narrative which oscillates between poles, but never settles into one (pure, authentic) place. Fanon continually moves position and moves on: in this sense, Fanon's work traces out a 'politics of position, movement and direction', where position is mapped through three (and more) spaces, where movement is perpetual, and direction is not a straight line toward some far off idealised objective. *Black Skin, White Masks* contributes to, and disrupts, a politics of cognitive mapping. It contributes by providing a spatial understanding of the place, and spaces, people occupy. It disrupts by ending any sense that a transparent map of real social relations is either possible or desirable. Sometimes, it is important to be invisible, unmappable (page 116); sometimes, it is important to be seen, to look back.

At second glance, Fanon's spaces are constitutive of the social relations within which bodies are, are marked and are imprisoned. In this reading, somehow, I managed to site Fanon's story on a street. I think it matters where this story took place; it matters that I misplaced Fanon. The significant point is that *where* this story is placed makes a difference to the 'politics' of the situation. I had assumed that Fanon was walking through a street in a colonial (Antillean) town. As he walked he was trying to meet the eyes/gaze of the Master as a deliberate act in the assertion of his equality. Nevertheless, he was having difficulty looking back and was eventually confronted by the screaming boy.

Some (rather obvious) clues disrupt this interpretation: (a) he is on a train, (b) there is snow on the ground and (c) he is in uniform. The story seems to be set in France: this would explain the train, the snow and the uniform (Fanon served with the French army in the Second World War). This setting would also explain the child's reaction. I have been puzzled as to why the child was so afraid if the encounter took place on a colonial street: was the white boy very sheltered so as to be surprised by the black man, or had he just stepped off the boat (as they say), or was Fanon walking down a street where black men are unexpected, or was the child afraid because Fanon 'caught his eye'? Each of these situations marks a different aspect of the catastrophic ambivalence of colonial urban life; each circumscribes the kinds of political practice that can be imagined or effected – to become more or less visible, to draw clearer or to disrupt clear boundaries.

It seems, to me, to be most likely that Fanon's encounter occurs on the Master's (white French) land. If this is so, then there are some specific implications for politics. To begin with, Fanon is 'out of place' (following Part 3 above). This is why the boy is shocked and afraid. Fanon is also in many spaces: one of which might be characterised as a 'third space' – Fanon is *both* 'in place' (in French uniform) *and* 'out of place' (black on white). He is neither on the margins nor at the centre, so the encounter provokes a series of 'third' moments. Fanon moves from Reason to Unreason to History (and Situation). Fanon moves from third person to first person to make a scene (at the expense of the white woman's desire). Although he seems to be caught in the racialised corporeal schema under the dissection of blue eyes, he struggles to look back. When he looks, he finds that he is trapped in a death dance with the other, where the steps have been marked out for them. They perform the rigid choreography of the Manichaean delirium: White and Black, Man and Woman, Beautiful and Ugly, Good and Bad, Close and Distant, Desire and Disgust, and so on. Fanon is moving toward the other, but he is not the fabled big bad ugly black man. He does, and does not, have a place in this macabre *commedia dell'arte*.

So far, I have stressed Fanon's representation of the fixity and fluidity, the movement from place to place, of the grotesque corporeal schema of colonial situations. Yet, his analysis has been taken to task for its dualistic and hierarchical imagination: because of this, he finds it impossible to understand certain positions, such as those of black women (see Doane, 1991) and homosexual blacks (see Mercer, 1995). These are significant exclusions, but Fanon's own use of dialectics works against their omission. The dialectic gives *Black Skin, White Masks* its structure – Black against White, Skin against Mask – and its content – the identifications and desires across lines of 'race' and 'sex'. But the ambivalent intersections of difference produce different, ambiguous differences. The content of categories of identity and desire are continuously thrown into question: white, black, mask, skin are neither pure nor authentic; no one is quite in the place they are seen, each changes in relation to a ghostly other.[19] One space: third spaces: other spaces. These are the shifting spaces of the politics of visibility, of misrecognition and of the *commedia dell'arte*. As Mercer says (1995: 20–21), 'the social field of fear and fantasy is never finally fixed. The oppressive regimes of myth and stereotype that inform the political management of multicultural discontent are themselves fluid, mobile and highly unpredictable, constantly updating themselves in the service of the changing same.'

The same changes, the other changes, as each occupies and moves through simultaneously real, imaginary and symbolic spatialities. No one is quite where they are seen to be. From this perspective, the politics of location or position must take into account not only the spaces of the subject, but also the ways in which people move through spaces which are constitutive of subjectivity, and also the dog-legged directions which any politics of subjectivity might involve. Moreover, it would be necessary to acknowledge the awkward reality of identification, desire and fantasy in psychic life. Otherwise, a disingenuous politics will align with an insidious 'Truth', both of which have been produced on the allegedly firm ground of 'dichotomous codifications of difference' (Mercer, 1995: 24). Instead, sustaining political communities will involve 'alliances and collaborations across divisive boundaries' and, despite internal rifts, a sense of 'spatial comradeship' (see Mohanty, 1991: 4).

A sense of political space might be established through cognitive mapping. But the cognitive map of real, imaginary and symbolic spatialities is not 'out there'; it will not provide definitive political practices, strategies and tactics. Instead, it will have much to do with the places where the lines of demarcation are drawn in the 'arbitrary closure' or 'strategic essentialism' of political mobilisation. On a flat space of position, it is possible to imagine secure boundaries which separate 'us' from 'them'. But in an unfolding space of subjectivity, the Swiss roll of politics is bound to be a little messier. Power relations are sticky, so 'communities of resistance' must be prepared to move in other 'third' spaces. This is not to abandon notions of 'position' or

---

19 For another sense of what this might mean, see Gilroy, 1993.

'arbitrary closure', but to suggest that 'spatial comradeship' lies in the continual process of way-finding, where the place you are heading is often (always?) just over the horizon, just out of sight.

And where are these places? Fanon's dialectics end up disrupting his analysis, crossed lines of identification and desire stretch out to positions he was not able to countenance. The cognitive map of powerful, certain political action crumbles into dust at the edges of the Empire of Reason. Maybe political positions are not to be discovered in the passive, fixed, undialectical space of absolute certainty, but in the place of psychodynamics.[20]

---

20 I wanted to conclude this book on the politics of listening, but I like the 'place of psychodynamics' exit too much. So Lacan's words will have to wait for another time: 'the art of listening is almost as important as saying the right thing' (Lacan, 1973: 123). I would prefer to delete the 'almost' though.

BIBLIOGRAPHY

Adorno, T. (1966) *Negative Dialectics*, 1990, London: Routledge.

———(1978) 'Freudian theory and the pattern of fascist propaganda', in T. Adorno, 1991, *The Culture Industry: Selected Essays on Mass Culture*, London: Routledge, 114–135.

Agnew, J. A. and Duncan, J. S. (1981) 'The transfer of ideas into Anglo-American human geography', *Progress in Human Geography* 5: 42–57.

Althusser, L. (1970) 'Ideology and ideological state apparatus', in L. Althusser, *Essays on Ideology*, 1984, London: Verso, 1–60.

Bachelard, G. (1957) *The Poetics of Space*, 1969, Boston: Beacon Press.

Barnes, T. and Duncan, T. (eds) (1992) *Writing Worlds: Discourse, Text and Metaphor in the Representation of Landscape*, London: Routledge.

Barthes, R. (1966) *Critique et Vérité*, Paris: Seuil.

Bégoin-Guignard, F. (1994) 'Comment', in S. Heald and A. Deluz (eds) *Anthropology and Psychoanalysis: An Encounter through Culture*, London: Routledge, 149–151.

Bell, D. and Valentine, G. (eds) (1995) *Mapping Desire: Geographies of Sexuality*, London: Routledge.

Benjamin, W. (1973) *Charles Baudelaire: A Lyric Poet in the Era of High Capitalism*, (1983, London: Verso.

Berger, J. (1972) *Ways of Seeing*, London and Harmondsworth: British Broadcasting Corporation and Penguin.

Bhabha, H. (1983) 'The other question: the stereotype and colonial discourse', in Screen Editorial (eds) *The Sexual Subject: A* Screen *Reader in Sexuality*, 1992, London: Routledge, 312–331.

———(1984) 'Of mimicry and man: the ambivalence of colonial discourse', in A. Michelson, R. Krauss, D. Crimp and J. Copjec (eds) *October: The First Decade, 1976–1986*, 1987, Cambridge, Massachusetts: MIT Press, 317–325.

———(1986) 'The other question: difference, discrimination and the discourse of colonialism', in R. Ferguson, M. Gever, M-h. T. Trinh and C. West (eds) *Out There: Marginalization and Contemporary Cultures*, 1990, Cambridge, Massachusetts: MIT Press, 71–87.

———(1990) 'Interrogating identity: Frantz Fanon and the postcolonial prerogative', in *The Location of Culture*, 1994, London: Routledge, 40–65.

———(1994) *The Location of Culture*, London: Routledge.

Billinge, M., Gregory, D. and Martin, R. L. (eds) (1984) *Recollections of a Revolution: Geography as Spatial Science*, London: Macmillan.

Blazwick, I., in conjunction with Francis, M., Wollen, P. and Imrie, M. (1989) *An Endless Adventure ... An Endless Passion ... An Endless Banquet: A Situationist Scrapbook*, London: Verso.

Blum, V. and Nast, H. (forthcoming) 'Where's the difference? The heterosexualization of alterity in Henri Lefebvre and Jacques Lacan', *Environment and Planning D: Society and Space*.

Bocock, R. (1976) *Freud and Modern Society: An Outline and Analysis of Freud's Sociology*, Wokingham: Van Nostrand Reinhold.

Boulding, K. E. (1956) *The Image: Knowledge in Life and Society*, Ann Arbor: University of Michigan Press.

———(1973) 'Foreword', in R. M. Downs and D. Stea (eds) *Image and Environment: Cognitive Mapping and Spatial Behaviour*, Chicago: Aldine, vii–xi.

Bowlby, R. (1993) *Shopping with Freud*, London: Routledge.

Brennan, T. (1992) *The Interpretation of the Flesh: Freud and Femininity*, London: Routledge.

Brookfield, H. C. (1969) 'On the environment as perceived', *Progress in Geography* 1: 51–80.
Brooks, K. (1973) 'Freudianism is not a basis for a Marxist psychology', in P. Brown (ed.) *Radical Psychology*, London: Tavistock, 315–374.
Brown, N. O. (1959) *Life against Death: The Psychoanalytical Meaning of History* (Second edition, 1985), Middletown, Connecticut: Wesleyan University Press.
Bruce Pratt, M. (1984) 'Identity: skin blood heart', in E. Bulkin, M. Bruce Pratt and B. Smith (eds) *Yours in Struggle: Three Feminist Perspectives on Anti-Semitism and Racism*, New York: Long Haul Press, 9–63.
Bruno, G. (1993) *Streetwalking on a Ruined Map: Cultural Theory and the City Films of Elvira Notari*, Princeton: Princeton University Press.
Buck-Morss, S. (1986) 'The flâneur, the sandwichman and the whore', *New German Critique* 39: 99–140.
——(1989) *The Dialectics of Seeing: Walter Benjamin and the Arcades Project*, Cambridge, Massachusetts: MIT Press.
Bunting, T. E. and Guelke, L. (1979) 'Behavioural and perception geography: a critical appraisal', *Annals of the Association of American Geographers* 69: 448–462.
Burgess, J., Limb, M. and Harrison, C. M. (1988a) 'Exploring environmental values through the medium of small groups. Part One: theory and practice', *Environment and Planning A* 20: 309–326.
——(1988b) 'Exploring environmental values through the medium of small groups. Part Two: illustrations of a group at work', *Environment and Planning A* 20: 457–476.
Burnett, P. (1976) 'Behavioural geography and the philosophy of mind', in R. G. Golledge and G. Rushton (eds) *Spatial Choice and Spatial Behaviour: Geographic Essays on the Analysis of Preferences and Perceptions*, Columbus: Ohio State University Press, 23–48.
Burton, I. (1963) 'The quantitative revolution and theoretical geography', *Canadian Geographer* 7: 151–162.
Butler, J. (1990) *Gender Trouble: Feminism and the Subversion of Identity*, New York: Routledge.
——(1993) *Bodies That Matter: On the Discursive Limits of 'Sex'*, New York: Routledge.
Buttimer, A. (1976) 'Grasping the dynamism of lifeworld', *Annals of the Association of American Geographers* 66: 277–292.
Callois, R. (1935) 'Mimicry and legendary psychasthenia', in A. Michelson, R. Krauss, D. Crimp and J. Copjec (eds) *October: The First Decade, 1976–1986*, 1987, translation dated 1984, Cambridge, Massachusetts: MIT Press; translation dated 1984, 58–74.
Cameron, D. (1985) *Feminism and Linguistic Theory*, London: Macmillan.
Carter, A. (1979) *The Bloody Chamber and Other Stories*, Harmondsworth: Penguin.
Chodorow, N. (1978) *The Reproduction of Mothering: Psychoanalysis and the Sociology of Gender*, Berkeley, California: University of California Press.
——(1989) *Feminism and Psychoanalytic Theory*, New Haven: Yale University Press.
Cixous, H. (1975) 'Sorties', in E. Marks and I. de Courtivon (eds) *New French Feminisms*, 1981, Sussex: Harvester Press, 90–98.
Cosgrove, D. (1982) 'The myth and the stories of Venice: an historical geography of a symbolic landscape', *Journal of Historical Geography* 8: 145–169.
——(1983) 'Towards a radical cultural geography: problems of theory', *Antipode* 15(1): 1–11.
——(1988) 'The geometry of landscape: practical and speculative art in sixteenth-century Venetian land territories', in D. Cosgrove and S. Daniels (eds) *The Iconography of Landscape: Essays on the Symbolic Representation, Design and Use of Past Environments*, Cambridge: Cambridge University Press, 254–276.
——(1989) 'Historical considerations on humanism, historical materialism and geography', in A. Kobayashi and S. Mackenzie (eds) *Remaking Human Geography*, London: Unwin Hyman, 189–205.
Couclelis, H. and Golledge, R. G. (1983) 'Analytic research, positivism and behavioural geography', *Annals of the Association of American Geographers* 73: 331–339.
Coward, R. and Ellis, J. (1977) *Language and Materialism*, London: Routledge and Kegan Paul.
Cox, K. R. (1981) 'Bourgeois thought and the behavioural geography debate', in K. Cox and R. G. Golledge (eds) *Behavioural Problems in Geography Revisited*, London: Methuen, 256–279.
Cox, K. R. and Golledge, R. G. (eds) (1969) *Behavioural Problems in Geography: A Symposium*, Evanston: Northwestern Studies in Geography, 17.
——(1981) 'Preface', in K. R. Cox and R. G. Golledge (eds) *Behavioural Problems in Geography Revisited*, London: Methuen, xiii–xxix.

Craib, I. (1989) *Psychoanalysis and Social Theory*, Hemel Hempstead: Harvester Wheatsheaf.
Creed, B. (1993) *The Monstrous-Feminine: Film, Feminism, Psychoanalysis*, London: Routledge.
Daniels, S. (1985) 'Arguments for a humanistic geography', in R. J. Johnston (ed.) *The Future of Geography*, London: Methuen, 143–158.
——(1989) 'Marxism, culture, and the duplicity of landscape', in R. Peet and N. Thrift (eds) *New Models in Geography: Volume 2*, London: Unwin Hyman, 196–220.
——(1993) *Fields of Vision: Landscape Imagery and National Identity in England and the United States*, Cambridge: Polity.
Dawson, G. (1994) *Soldier Heroes: British Adventure, Empire and the Imagining of Masculinities*, London: Routledge.
de Beauvoir, S. (1949) *The Second Sex*, 1953, London: Jonathan Cape.
de Certeau, M. (1984) *The Practice of Everyday Life*, London: University of California Press.
de Lauretis, T. (1994) *The Practice of Love: Lesbian Sexuality and Perverse Desire*, Bloomington: Indiana University Press.
Debord, G. (1967) *Society of the Spectacle*, 1987, London: Rebel Press.
Deleuze, G. and Guattari, F. (1972) *Anti-Oedipus: Capitalism and Schizophrenia*, 1984, London: Athlone.
——(1987) *A Thousand Plateaus: Capitalism and Schizophrenia*, 1988, London: Athlone.
Deutsch, H. (1930) 'The significance of masochism in the mental life of women', *International Journal of Psycho-Analysis* 9: 48–60.
Doane, M. A. (1991) 'Dark continents: epistemologies of racial and sexual difference in psychoanalysis and the cinema', in *Femmes Fatales: Feminism, Film Theory, Psychoanalysis*, London: Routledge, 209–248.
Domosh, M. (1992) 'Corporate cultures and the modern landscape of New York City', in K. Anderson and F. Gale (eds) *Inventing Places: Studies in Cultural Geography*, London: Longman, 72–88.
Douglas, M. (1966) *Purity and Danger: An Analysis of the Concepts of Pollution and Taboo*, London: Routledge.
Downs, R. M. (1970) 'Geographic space perception: past approaches and future prospects', *Progress in Geography* 2: 65–108.
——(1979) 'Critical appraisal or determined philosophical skepticism? Commentary', *Annals of the Association of American Geographers* 69: 463–464.
Downs, R. M. and Meyer, J. T. (1978) 'Geography and the mind: an exploration of perceptual geography', *American Behavioural Scientist* 22(1): 59–77.
Downs, R. M. and Stea, D. (1973a) 'Cognitive maps and spatial behaviours: processes and products', in R. M. Downs and D. Stea (eds) *Image and Environment: Cognitive Mapping and Spatial Behaviour*, Chicago: Aldine, 8–26.
——(1973b) 'Introduction', in R. M. Downs and D. Stea (eds) *Image and Environment: Cognitive Mapping and Spatial Behaviour*, Chicago: Aldine, 1–7.
——(1977) *Maps in Mind: Reflections on Cognitive Mapping*, London: Harper and Row.
Driver, F. (1985) 'Power, space and the body: a critical assessment of Foucault's *Discipline and Punish*', *Environment and Planning D: Society and Space* 3: 425–446.
——(1988) 'Moral geographies: social science and the urban environment in mid-nineteenth century England', *Transactions of the Institute of British Geographers* 13: 275–287.
——(1992) 'Geography's empire: histories of geographical knowledge', *Environment and Planning D: Society and Space* 10(1): 23–40.
Duncan, J. (ed.) (1982) *Housing and Identity*, London: Croom Helm.
Duncan, J. and Ley, D. (1982) 'Structural Marxism and human geography: a critical assessment', *Annals of the Association of American Geographers* 72: 30–59.
——(eds) (1993) *Place/Culture/Representation*, London: Routledge.
Duncan, J. S. and Duncan, N. G. (1976) 'Housing as presentation of self and the structure of social networks', in G. T. Moore and R. G. Golledge (eds) *Environmental Knowing*, London: Hutchinson, 247–253.
Elias, N. (1939) *The Civilizing Process. Volume 1: The History of Manners*, 1978, New York: Pantheon.
Elliott, A. (1992) *Social Theory and Psychoanalysis in Transition: Self and Society from Freud to Kristeva*, Oxford: Basil Blackwell.
——(1994) *Psychoanalytic Theory: An Introduction*, Oxford: Basil Blackwell.
Engels, F. (1844) *The Condition of the Working Class*, 1969, Harmondsworth: Panther.
Erikson, E. H. (1950) *Childhood and Society*, New York: W. W. Norton.
——(1959) *Identity and the Life Cycle*, New York: International Universities Press.

Eyles, J. (1989) 'The geography of everyday life', in D. Gregory and R. Walford (eds) *Horizons in Human Geography*, London: Macmillan, 102–117.
Fanon, F. (1952) *Black Skin, White Masks*, 1986, London: Pluto Press.
Firestone, S. (1970) *The Dialectic of Sex: The Case for Feminist Revolution*, 1979, London: Women's Press.
Fisher, C. (1956) 'Dreams, images and perception: a study of unconscious–preconscious relationships', *Journal of the American Psychoanalytic Association* 4: 5–48.
Fisher, C. and Paul, I. H. (1959) 'The effect of subliminal visual stimulation on images and dreams: a validation study', *Journal of the American Psychoanalytic Association* 7: 35–83.
Fletcher, J. and Benjamin, A. (1990) *Abjection, Melancholia and Love: The Work of Julia Kristeva*, London: Routledge.
Foucault, M. (1961) *Madness and Civilization: A History of Insanity in the Age of Reason*, 1967, London: Routledge.
——(1966) *The Order of Things: An Archaeology of the Human Sciences*, 1970, London: Routledge.
Foulkes, S. and Anthony, E. (1957) *Group Psychotherapy: The Psycho-analytic Approach*, Harmondsworth: Penguin.
Freud, S. (1900) *The Interpretation of Dreams*, 1976, Volume 4, Harmondsworth: Penguin Freud Library.
——(1901) *The Psychopathology of Everyday Life*, 1975, Volume 5, Harmondsworth: Penguin Freud Library.
——(1905) 'Three essays on the theory of sexuality', in *On Sexuality: Three Essays on the Theory of Sexuality and Other Works*, 1977, Volume 7, Harmondsworth: Penguin Freud Library, 39–169.
——(1908) 'On the sexual theories of children', in *On Sexuality: Three Essays on the Theory of Sexuality and Other Works*, 1977, Volume 7, Harmondsworth: Penguin Freud Library, 187–204.
——(1911) 'Formulations on the two principles of mental functioning', in *On Metapsychology: The Theory of Psychoanalysis*, 1984, Volume 11, Harmondsworth: Penguin Freud Library, 35–44.
——(1913) 'Totem and taboo', in *The Origins of Religion*, 1985, Volume 13, Harmondsworth: Penguin Freud Library, 49–224.
——(1914) 'On narcissism: an introduction', in *On Metapsychology: The Theory of Psychoanalysis*, 1984, Volume 11, Harmondsworth: Penguin Freud Library, 65–97.
——(1915a) 'The unconscious', in *On Metapsychology: The Theory of Psychoanalysis*, 1984, Volume 11, Harmondsworth: Penguin Freud Library, 167–222.
——(1915b) 'Instincts and their vicissitudes', in *On Metapsychology: The Theory of Psychoanalysis*, 1984, Volume 11, Harmondsworth: Penguin Freud Library, 113–138.
——(1915c) 'Repression', in *On Metapsychology: The Theory of Psychoanalysis*, 1984, Volume 11, Harmondsworth: Penguin Freud Library, 145–158.
——(1917a) 'A difficulty in the path of psychoanalysis', in S. Freud, *Standard Edition of the Complete Works of Sigmund Freud*, 1927–1931, Volume XVII, London: Hogarth Press.
——(1917b) 'Mourning and melancholia', in *On Metapsychology: The Theory of Psychoanalysis*, 1984, Volume 11, Harmondsworth: Penguin Freud Library, 251–268.
——(1919) 'The "uncanny"', in *Art and Literature: Jensen's 'Gradiva', Leonardo Da Vinci and other works*, 1985, Volume 14, Harmondsworth: Penguin Freud Library, 339–376.
——(1920) 'Beyond the pleasure principle', in *On Metapsychology: The Theory of Psychoanalysis*, 1984, Volume 11, Harmondsworth: Penguin Freud Library, 275–338.
——(1921) 'Group psychology and the analysis of the ego', in *Civilization, Society and Religion: Group Psychology, Civilization and its Discontents and Other Works*, 1985, Volume 12, Harmondsworth: Penguin Freud Library, 95–178.
——(1923) 'The ego and the id', in *On Metapsychology: The Theory of Psychoanalysis*, 1984, Volume 11, Harmondsworth: Penguin Freud Library, 350–407.
——(1925) 'Some psychical consequences of the anatomical distinction between the sexes', in *On Sexuality: Three Essays on the Theory of Sexuality and Other Works*, 1977, Volume 7, Harmondsworth: Penguin Freud Library, 331–343.
——(1926a) 'The question of lay-analysis: conversations with an impartial person', in S. Freud, *Two Short Accounts of Psycho-analysis*, 1962, Harmondsworth: Pelican, 91–170.
——(1926b) 'Inhibitions, symptoms and anxiety', in *On Psychopathology: Inhibitions, Symptoms and Anxiety and Other Works*, 1979, Volume 10, Harmondsworth: Penguin Freud Library, 237–333.

———(1927) 'The future of an illusion', in *Civilization, Society and Religion: Group Psychology, Civilization and its Discontents and Other Works*, 1985, Volume 12, Harmondsworth: Penguin Freud Library, 183–241.

———(1930) 'Civilization and its discontents', in *Civilization, Society and Religion: Group Psychology, Civilization and its Discontents and Other Works*, 1985, Volume 12, Harmondsworth: Penguin Freud Library, 251–340.

———(1933) *New Introductory Lectures on Psychoanalysis*, 1973, Volume 2, Harmondsworth: Penguin Freud Library.

———(1939) 'Moses and monotheism: three essays', in *The Origins of Religion*, 1985, Volume 13, Harmondsworth: Penguin Freud Library, 243–386.

———(1940) 'Splitting of the ego in the process of defence', in *On Metapsychology: The Theory of Psychoanalysis*, 1984, Volume 11, Harmondsworth: Penguin Freud Library, 461–464.

Fromm, E. (1973) *The Anatomy of Human Destructiveness*, 1977, Harmondsworth: Penguin.

Frosh, S. (1987) *The Politics of Psychoanalysis: An Introduction to Freudian and Post-Freudian Theory*, Basingstoke: Macmillan.

Fuss, D. (1995) *Identification Papers: Reflections on Psychoanalysis, Sexuality and Culture*, London: Routledge.

Gallop, J. (1982) *Feminism and Psychoanalysis: The Daughter's Seduction*, London: Macmillan.

———(1988) *Thinking Through the Body*, New York: Columbia University Press.

Gay, P. (1984) *The Bourgeois Experience: Victoria to Freud. Volume 1: The Education of the Senses*, New York: Oxford University Press.

———(1988) *Freud: A Life for our Time*, London: Dent.

Getis, A. and Boots, B. N. (1971) 'Spatial behaviour: rats and man', *Professional Geographer* 23(1): 11–14.

Giddens, A. (1976) *New Rules of Sociological Method: Positive Critique of Interpretative Sociologies*, London: Hutchison.

———(1979) *Central Problems in Social Theory: Action, Structure and Contradiction in Social Analyses*, London: Macmillan.

———(1984) *The Constitution of Society: Outline of the Theory of Structuration*, Cambridge: Polity Press.

———(1991) *Modernity and Self-Identity*, Cambridge: Polity Press.

———(1992) *The Transformation of Intimacy: Sexuality, Love and Eroticism in Modern Societies*, Cambridge: Polity Press.

Gilroy, P. (1993) *The Black Atlantic: Modernity and Double Consciousness*, London: Verso.

Goffman, E. (1959) *The Presentation of Self in Everyday Life*, New York: Anchor Books.

Gold, J. (1980) *An Introduction to Behavioural Geography*, Oxford: Oxford University Press.

Gold, J. and Goody, B. (1984) 'Behavioural geography and perceptual geography: criticisms and responses', *Progress in Human Geography* 8: 544–550.

Golledge, R. G. (1967) 'Conceptualizing the market decision process', *Journal of Regional Science* 7(2) (supplement): 239–258.

———(1981) 'Misconceptions, misinterpretations and misrepresentations of behavioural approaches in human geography', *Environment and Planning A* 13: 1325–1344.

Golledge, R. G. and Rushton, G. (1976) 'Introduction', in R. G. Golledge and G. Rushton (eds) *Spatial Choice and Spatial Behaviour*, Columbus: Ohio State University Press, 1–2.

Golledge, R. G., Brown, L. A. and Williamson, F. (1972) 'Behavioural approaches in geography: an overview', *Australian Geographer* 12: 59–79.

Gould, P. (1973) 'On mental maps', in R. M. Downs and D. Stea (eds) *Image and Environment: Cognitive Mapping and Spatial Behaviour*, Chicago: Aldine, 182–220.

———(1976) 'Cultivating the garden: a commentary and critique on some multi-dimensional speculations', in R. G. Golledge and G. Rushton (eds) *Spatial Choice and Spatial Behaviour*, Columbus: Ohio State University Press, 83–91.

Gould, P. and White, R. (1974) *Mental Maps*, 1986, London: George Allen and Unwin.

Gregory, D. (1978) *Ideology, Science and Human Geography*, London: Hutchinson.

———(1981) 'Human agency and human geography', *Transactions of the Institute of British Geographers* 6: 1–18.

———(1989a) 'The crisis of modernity? Human geography and critical social theory', in R. Peet and N. Thrift (eds) *New Models in Geography: Volume 2*, London: Unwin Hyman, 348–385.

———(1989b) 'Areal differentiation and post-modern human geography', in D. Gregory and R. Walford (eds) *Horizons in Human Geography*, London: Macmillan, 67–96.

———(1994) *Geographical Imaginations*, Oxford: Basil Blackwell.

——(1995) 'Lefebvre, Lacan and the production of space' in G. Benko and U. Strohmeyer (eds) *Geography, History and Social Science*, Dordrecht: Kluwer.
Gregson N. (1986) 'On duality and dualism: the case of time-geography and structuration', *Progress in Human Geography* 10: 184–205.
Grosz, E. (1989) *Sexual Subversions: Three French Feminists*, London: Routledge.
——(1990) *Jacques Lacan: A Feminist Introduction*, London: Routledge.
——(1994) *Volatile Bodies: Toward a Corporeal Feminism*, Bloomington: Indiana University Press.
Habermas, J. (1972) *Knowledge and Human Interests*, London: Heinemann.
——(1973) *Legitimation Crisis*, 1976, London: Heinemann.
Hägerstrand, T. (1965) 'A Monte Carlo approach to diffusion', *Archives Européenes de Sociologie* 6(1): 43–67.
——(1970) 'What about people in regional science?', *Papers and Proceedings of the Regional Science Association* 24: 7–21.
Haraway, D. (1989) *Primate Visions: Gender, Race, and Nature in the World of Modern Science*, London: Routledge.
——(1990) 'A manifesto for cyborgs: science, technology and socialist feminism in the 1980s', in L. Nicholson (ed.) *Feminism/Postmodernism*, London: Routledge, 190–233.
——(1991) *Simians, Cyborgs and Women: The Reinvention of Nature*, London: Free Association Books.
Harvey, D. (1966) 'Geographical process and the analysis of point patterns', *Transactions of the Institute of British Geographers* 40: 81–95.
——(1969a) *Explanation in Geography*, London: Edward Arnold.
——(1969b) 'Conceptual and measurement problems in the cognitive-behavioural approach to location theory', in K. R. Cox and R. G. Golledge (eds) *Behavioural Problems in Geography Revisited*, 1969, London: Methuen, 18–42.
——(1985) *The Urbanization of Capital*, Oxford: Basil Blackwell.
Harvey, M. and Holly, B. (eds) (1981) *Themes in Geographic Thought*, London: Croom Helm.
Heald, S. and Deluz, A. (eds) (1994) *Anthropology and Psychoanalysis: An Encounter through Culture*, London: Routledge.
Held, D. (1980) *Introduction to Critical Theory: Horkheimer to Habermas*, Hutchinson: London.
Horkheimer, M. and Adorno, T. (1944) *Dialectic of Enlightenment*, 1972, London: Verso.
Horney, K. (1932) 'The dread of women: observation on a specific difference in the dread felt by men and by women respectively for the opposite sex', *International Journal of Psycho-Analysis* 13: 348–360.
——(1933) 'The denial of the vagina: a contribution to the problem of the genital anxieties specific to women', in K. Horney, *Feminine Psychology*, New York: W. W. Norton, 147–161.
Hudson, R. (1980) 'Personal construct theory: the repertory grid method and human geography', *Progress in Human Geography* 4: 346–359.
Huff, D. L. (1960) 'A topographic model of consumer space preferences', *Papers and Proceedings of the Regional Science Association* 6: 159–173.
Huxley, A. (1956) *Heaven and Hell*, 1977, London: Grafton.
Huxtable, A. L. (1984) *The Tall Building Artistically Reconsidered: The Search for a Skyscraper Style*, New York, Pantheon Books.
Irigaray, L. (1974) *Speculum of the Other Woman*, 1985, Ithaca, New York: Cornell University Press.
——(1977) *This Sex Which is Not One*, 1985, Ithaca, New York: Cornell University Press.
——(1993) *An Ethics of Sexual Difference*, London: Athlone Press.
Jackson, P. (1989) *Maps of Meaning: An Introduction to Cultural Geography*, London: Routledge.
Jacobs, J. (1961) *The Death and Life of Great American Cities*, New York: Vintage.
Jameson, F. (1984) 'Postmodernism, or the cultural logic of late capitalism', *New Left Review* 146: 53–92.
——(1991) *Postmodernism, or the Cultural Logic of Late Capitalism*, London: Verso.
Jay, M. (1993) *Downcast Eyes: The Denigration of Vision in Twentieth Century French Thought*, Berkeley: University of California Press.
Johnston, R. J. (1986) *Philosophy and Human Geography: An Introduction to Contemporary Approaches* (Second edition), London: Edward Arnold.
——(1987) *Geography and Geographers: Anglo-American Human Geography since 1945* (Third edition), London: Edward Arnold.

Jones, E. (1927) 'The early development of female sexuality', *International Journal of Psycho-Analysis* 8: 459–472.
Keith, M. (1987) '"Something Happened": the problems of explaining the 1980 and 1981 riots in British cities', in P. Jackson (ed.) *Race and Racism: Essays in Social Geography*, London: Allen and Unwin, 275–325.
Keith, M. and Pile, S. (1993a) 'Introduction: Part 1. The politics of place ...', in M. Keith and S. Pile (eds) *Place and the Politics of Identity*, London: Routledge, 1–21.
——(eds) (1993b) *Place and the Politics of Identity*, London: Routledge.
Kelly, G. A. (1955) *The Psychology of Personal Constructs*, New York: W. W. Norton.
Kirby, K. (1996) *Indifferent Boundaries: Exploring the Space of the Subject*, New York: Guilford Press.
Kirk, W. (1952) 'Historical geography and the concept of the behavioural environment', *Indian Geographical Journal* 25: 152–160.
——(1963) 'Problems of geography', *Geography* 48: 357–371.
Klein, M. (1927) 'Symposium on child-analysis', in M. Klein, *Love, Guilt and Reparation, and Other Works, 1921–1945*, 1988, London: Virago, 139–169.
——(1945) 'The Oedipus complex in the light of early anxieties', in M. Klein, *Love, Guilt and Reparation, and Other Works, 1921–1945*, 1988, London: Virago, 370–419.
——(1959) 'Our adult world and its roots in infancy', in M. Klein, *Envy and Gratitude, and Other Works: 1946–1963*, 1988, London: Virago, 247–263.
Knapp, P. H. (1956) 'Sensory impressions in dreams', *Psychoanalytic Quarterly* 25: 325–347.
Kobayashi, A. and Mackenzie, S. (eds) (1989) *Remaking Human Geography*, London: Unwin Hyman.
Koffka, K. (1929) *Principles of Gestalt Psychology*, 1952, New York: Harcourt-Brace.
Köhler, W. (1929) *Gestalt Psychology*, 1947, New York: Harcourt-Brace.
Kolodny, A. (1973) *The Lay of the Land: Metaphor as Experience and History in American Life and Letters*, Chapel Hill: University of North Carolina Press.
Kristeva, J. (1977) *Desire in Language: A Semiotic Approach to Literature and Art*, 1980, Oxford: Basil Blackwell.
——(1980) *Powers of Horror: An Essay on Abjection*, 1982, New York: Columbia University Press.
Lacan, J. (1949) 'The mirror stage as formative of the function of the I', in J. Lacan, *Écrits: A Selection*, 1977, London: Tavistock, 1–7.
——(1953) 'The function and field of speech and language in psychoanalysis', in J. Lacan, *Écrits: A Selection*, 1977, London: Tavistock, 30–113.
——(1958) 'The signification of the phallus', in J. Lacan, *Écrits: A Selection*, 1977, London: Tavistock Publications, 281–291. An alternative translation is published in J. Mitchell and J. Rose (eds) *Feminine Sexuality*, 1982, London: Macmillan, 74–85.
——(1959) 'On a question preliminary to any possible treatment of psychosis', in J. Lacan, *Écrits: A Selection*, 1977, London: Tavistock, 179–225.
——(1972–3) 'God and the *jouissance* of ~~The~~ woman', in J. Mitchell and J. Rose (eds) *Feminine Sexuality: Jacques Lacan and the* Ecole Freudienne, 1982, London: Macmillan, 138–148.
——(1973) *The Four Fundamental Concepts of Psycho-analysis*, 1986, Harmondsworth: Peregrine Books.
Laplanche, J. and Pontalis, J. B. (1973) *The Language of Psychoanalysis*, 1988, London: Karnac Books and the Institute of Psycho-analysis.
Law, L. (forthcoming) 'A matter of "choice": discourses on prostitution in the Philippines', in L. Manderson and M. Jolly (eds) *Sites of Desire/Economies of Pleasure: Sexualities in Asia and the Pacific*, Chicago: University of Chicago Press.
Lebeau, V. (1995) *Lost Angels: Psychoanalysis and Cinema*, London: Routledge.
Lefebvre, H. (1974) *The Production of Space*, 1991, Oxford: Basil Blackwell.
Lévi-Strauss, C. (1949) *The Elementary Structures of Kinship*, 1969, Boston, Mass.: Beacon Press.
Lewis, C. and Pile, S. (1996) 'Woman, body, space: Rio Carnival and the politics of performance', *Gender, Place and Culture* 3(1): 23–41.
Ley, D. (1974) *The Black Inner City as Frontier Outpost: Images and Behavior of a Philadelphia Neighborhood*, Washington, Monograph Series number 7: Association of American Geographers.
——(1977) 'Social geography and the taken for granted world', *Transactions of the Institute of British Geographers* 2: 498–512.
——(1978) 'Social geography and social action', in D. Ley and M. S. Samuels (eds) *Humanistic*

Geography: Prospects and Problems, London: Croom Helm, 41–57.
——(1981a) 'Behavioural geography and the philosophies of meaning', in G. R. Cox and R. G. Golledge (eds) *Behavioural Problems in Geography Revisited*, London: Methuen, 209–230.
——(1981b) 'Cultural/humanistic geography', *Progress in Human Geography* 5: 249–257.
——(1982) 'Rediscovering man's place', *Transactions of the Institute of British Geographers* 7: 248–253.
——(1989) 'Fragmentation, coherence, and the limits to theory in human geography', in A. Kobayashi and S. Mackenzie (eds) *Remaking Human Geography*, London: Unwin Hyman, 227–244.
Ley, D. and Samuels, M. S. (1978a) 'Introduction: contexts of modern humanism in geography', in D. Ley and M. S. Samuels (eds) *Humanistic Geography: Prospects and Problems*, London: Croom Helm, 1–17.
——(1978b) *Humanistic Geography: Prospects and Problems*, London: Croom Helm.
Little, J., Peake, L. and Richardson, P. (eds) (1988) *Women in Cities: Gender and the Urban Environment*, London: Macmillan Education.
Livingstone, D. (1992) *The Geographical Tradition: Episodes in the History of a Contested Enterprise*, Oxford: Basil Blackwell.
Lloyd, G. (1993) *The Man of Reason: 'Male' and 'Female' in Western Philosophy* (Second edition), London: Routledge.
Longhurst, R. (1995) 'The body and geography', *Gender, Place and Culture* 2(1): 97–105.
Lowenthal, D. (1961) 'Geography, experience and imagination: towards a geographical epistemology', *Annals of the Association of American Geographers* 51: 241–260.
Lynch, K. (1960) *The Image of the City*, Cambridge, Massachusetts: MIT Press.
——(1976) 'Foreword', in G. T. Moore and R. G. Golledge (eds) *Environmental Knowing*, London: Hutchinson, v–viii.
MacCannell, J. F. (1991) *The Regime of the Brother: After the Patriarchy*, London: Routledge.
McDowell, L. and Court, L. (1994) 'Performing work: bodily representations in merchant banks', *Environment and Planning D: Society and Space* 12(6): 727–750.
Mahler, M. (1972) *The Psychological Birth of the Human Infant: Symbiosis and Individuation*, 1975, New York: Basic Books.
Marcuse, H. (1956) *Eros and Civilization: A Philosophical Inquiry into Freud*, 1987, London: Ark.
——(1964) *One Dimensional Man: Studies in the Ideology of Advanced Industrial Society*, 1972, London: Abacus.
Marx, K. (1852) 'The Eighteenth Brumaire of Louis Bonaparte', in K. Marx, *Surveys from Exile*, 1973, Harmondsworth: Penguin, 143–249.
Maslow, A. (1954) *Motivation and Personality*, New York: Harper and Row.
Massey, D. (1975) 'Behavioural research', *Area* 7: 210–213.
——(1994) *Space, Place and Gender*, Cambridge: Polity Press.
——(1995) 'Imagining the world', in J. Allen and D. Massey (eds) *Geographical Worlds*, Oxford: Oxford University Press/Open University, 5–52.
Matrix. (1984) *Making Space: Women and the Man-made Environment*, London: Pluto.
Mead, G. H. (1934) *Mind, Self and Society: From the Standpoint of a Social Behaviourist*, Chicago: Chicago University Press.
Mercer, D. (1972) 'Behavioural geography and the sociology of social action', *Area* 4: 48–52.
Mercer, K. (1989) 'Skin head sex thing: racial difference and the homoerotic imaginary', in *Welcome to the Jungle: New Positions in Black Cultural Studies*, 1994, London: Routledge, 189–219.
——(1994) *Welcome to the Jungle: New Positions in Black Cultural Studies*, London: Routledge.
——(1995) 'Busy in the ruins of wretched phantasia', in R. Farr (ed.) *Mirage: Enigmas of Race, Difference and Desire*, London: Institute of Contemporary Arts/Institute of International Visual Culture, 12–55.
Merleau-Ponty, M. (1945) *The Phenomenology of Perception*, 1964, Evanston: Northwestern University Press.
——(1964) *The Visible and the Invisible*, 1969, Evanston: Northwestern University Press.
Mitchell, J. (1974) *Psychoanalysis and Feminism: A Radical Reassessment of Freudian Psychoanalysis*, Harmondsworth: Penguin.
——(1982) 'Introduction – 1', in J. Mitchell and J. Rose (eds) *Feminine Sexuality: Jacques Lacan and the* Ecole Freudienne, London: Macmillan, 1–26.

Mohanty, C. T. (1991) 'Cartographies of struggle', in C. T. Mohanty, A. Russo and L. Torres (eds) *Third World Women and the Politics of Feminism*, Bloomington: Indiana University Press, 1–47.

Money-Kyrle, R. E. (1956) 'The world of the unconscious and the world of commonsense', *British Journal for the Philosophy of Science* 7: 86–96.

———(1960) *Man's Picture of His World: A Psycho-analytic Perspective*, London: Duckworth.

Monk, J. (1992) 'Gender in the landscape: expressions of power and meaning', in K. Anderson and F. Gale (eds) *Inventing Places: Studies in Cultural Geography*, London: Longman, 123–138.

Montrelay, M. (1970) 'Inquiry into femininity', in P. Adams and E. Cowie (eds) *The Woman Question: m/f*, 1990, London: Verso, 253–273.

Moore, H. (1994) 'Gendered persons: dialogues between anthropology and psychoanalysis', in S. Heald and A. Deluz (eds) *Anthropology and Psychoanalysis: An Encounter through Culture*, London: Routledge, 131–149.

Morris, M. (1992) 'Great moments in social climbing: King Kong and the Human Fly', in B. Colomina (ed.) *Sexuality and Space*, Princeton: Princeton Architectural Press, 1–51.

Mort, F. (1987) *Dangerous Sexualities: Medico-moral Politics in England since 1930*, London: Routledge.

Mulvey, L. (1975) 'Visual pleasure and narrative cinema', in Screen Editorial (eds) *The Sexual Subject: A* Screen *Reader in Sexuality*, 1992, London: Routledge, 22–34.

Mumford, L. (1961) *The City in History*, Harmondsworth: Penguin.

Munt, S. (1994) 'The lesbian flâneur', *Perversions* 1 (Winter): 61–81.

Nead, L. (1992) *The Female Nude: Art, Obscenity and Sexuality*, London: Routledge.

Ogborn, M. (1992) 'Love-state-ego: "centres" and "margins" in 19th century Britain', *Environment and Planning D: Society and Space* 10: 287–305.

Olivier, C. (1980) *Jocasta's Children: The Imprint of the Mother* (English translation 1989), Routledge, London.

Olsson, G. (1969) 'Inference problems in locational analysis', in K. R. Cox and R. G. Golledge (eds) *Behavioural Problems in Geography: A Symposium*, Evanston, Northwestern Studies in Geography 17, 14–34.

———(1980) *Birds In Egg/Eggs In Bird*, London: Pion.

———(1991) *Lines of Power: Limits of Language*, Minneapolis: University of Minnesota Press.

Park, R. E. (1925) 'The city: suggestions for investigation of human behavior in the urban environment', in R. E. Park and E. W. Burgess with R. D. McKenzie and L. Wirth, *The City: Suggestions for Investigation of Human Behavior in the Urban Environment*, 1984, Chicago: Midway Reprint, University of Chicago Press, 1–46.

Pederson-Krag, G. (1956) 'The use of metaphor in analytic thinking', *Psychoanalytic Quarterly* 25: 66–71.

Peet, R. (ed.) (1977) *Radical Geography: Alternative Viewpoints on Contemporary Social Issues*, London: Methuen.

Perin, C. (1988) *Belonging in America*, Madison: University of Wisconsin Press.

Pickles, J. (1985) *Phenomenology, Science and Geography: Spatiality and the Human Sciences*, Cambridge: Cambridge University Press.

Pile, S. (1990) 'Depth hermeneutics and critical human geography', *Environment and Planning D: Society and Space* 8(2): 211–232.

———(1991) 'Practising interpretative geography', *Transactions of the Institute of British Geographers* 16: 458–469.

———(1994a) 'Echo, desire and the grounds of knowledge: a mytho-poetic assessment of Buttimer's *Geography and the Human Spirit*', *Environment and Planning D: Society and Space* 12: 495–507.

———(1994b) 'Masculinism, the use of dualistic epistemologies, and third spaces', *Antipode* 26: 255–277.

Pile, S. and Thrift, N. (eds) (1995a) *Mapping the Subject: Geographies of Cultural Transformation*, London: Routledge.

———(1995b) 'Mapping the subject', in S. Pile and N. Thrift (eds) *Mapping the Subject: Geographies of Cultural Transformation*, London: Routledge, 13–51.

Pocock, D. C. D. (1973) 'Environmental perception: process and product', *Tijdschrift voor Economische en Sociale Geografie* 63: 251–257.

Pocock, D. C. D. and Hudson, R. (1978) *Images of the Urban Environment*, London: Macmillan.

Pollock, G. (1988a) *Vision and Difference: Femininity, Feminism and the Histories of Art*. London: Routledge.
———(1988b) 'Modernity and the spaces of femininity', in G. Pollock, *Vision and Difference: Femininity, Feminism and the Histories of Art*, London: Routledge, 50–90.
Porteus, J. D. (1977) *Environment and Behaviour: Planning and Everyday Urban Life*, Reading, Massachusetts: Addison-Wesley.
Poster, M. (1978) *Critical Theory of the Family*, London: Pluto.
Pratt, G. (1992) 'Spatial metaphors and speaking positions', *Environment and Planning D: Society and Space* 10: 241–244.
Pred, A. (1967) *Behaviour and Location: Foundations for a Geographic Dynamic Location Theory: Part 1*, Lund: Lund Studies in Geography, Series B, 27.
Probyn, E. (1993) *Sexing the Self: Gendered Positions in Cultural Studies*, London: Routledge.
Reich, W. (1946) *The Mass Psychology of Fascism*, Harmondsworth: Penguin.
Ricoeur, P. (1970) *Freud and Philosophy: An Essay on Interpretation*, London: Yale University Press.
———(1974) *The Conflict of Interpretations*, Evanston: Northwestern University Press.
Rieff, P. (1959) *Freud: The Mind of the Moralist*, 1979, Chicago: University of Chicago Press.
Riviere, J. (1929) 'Womanliness as masquerade', in V. Burgin, J. Donald and C. Kaplan (eds) *Formations of Fantasy*, 1986, London: Methuen, 35–44.
Rock, P. (1979) *The Making of Symbolic Interactionism*, London: Macmillan.
Rose, G. (1991) 'On being ambivalent: women and feminisms in geography', in C. Philo (compiler) *New Words, New Worlds: Reconceptualising Social and Cultural Geography. Conference Proceedings*, Lampeter: Department of Geography, St David's University College, 78–87.
———(1993) *Feminism and Geography: The Limits of Geographical Knowledge*, Cambridge: Polity Press.
———(1995) 'As if the mirrors had bled: masculine dwelling, masculinist theory and feminist masquerade', in N. Duncan (ed.) *BodySpaces*, London: Routledge.
Rose, J. (1982) 'Introduction – II', in J. Mitchell and J. Rose (eds) *Feminine Sexuality: Jacques Lacan and the* Ecole Freudienne, London: Macmillan, 27–58.
———(1986) *Sexuality in the Field of Vision*, London: Verso.
Rox, G. (1995) 'Distance, surface, elsewhere: a feminist critique of the space of phallocentric self/knowledge', *Environment and Planning D: Society and Space* 13: 761–781.
Rushton, G. (1969) 'Analysis of behaviour by revealed space preference', *Annals of the Association of American Geographers* 59: 391–400.
———(1979) 'Commentary on "Behavioural and perception geography"', *Annals of the Association of American Geographers* 69: 463–464.
Russo, M. (1994) *The Female Grotesque: Risk, Excess and Modernity*, New York: Routledge.
Saarinen, T. F. (1979) 'Commentary on "Behavioural and perception geography"', *Annals of the Association of American Geographers* 69: 464–468.
Said, E. (1978) *Orientalism*, Harmondsworth: Peregrine.
———(1993) *Culture and Imperialism*, London: Vintage.
Samuels, R. (1993) *Between Philosophy and Psychoanalysis: Lacan's Reconstruction of Freud*, London: Routledge.
Sartre, J.-P. (1943) *Being and Nothingness: An Essay on Phenomenological Ontology*, 1958, London: Routledge.
Sauer, C. O. (1941) 'Foreword to historical geography', *Annals of the Association of American Geographers* 31: 1–24.
Sayer, A. (1989) 'On the dialogue between humanism and historical materialism in geography', in A. Kobayashi and S. Mackenzie (eds) *Remaking Human Geography*, London: Unwin Hyman, 206–226.
Schor, N. (1992) 'Feminism and George Sand: *Lettres à Marcie*', in J. Butler and J. W. Scott (eds) *Feminists Theorize the Political*, New York: Routledge, 41–53.
Seamon, D. (1979) *A Geography of the Lifeworld: Movement, Rest, and Encounter*, London: Croom Helm.
———(1980) 'Body-subject, time-space routines, and place-ballets', in A. Buttimer and D. Seamon (eds) *The Human Experience of Space and Place*, London: Croom Helm, 148–165.
Segal, N. (1988) *Narcissus and Echo: Women in the French Récit*, Manchester, Manchester University Press.
Sennett, R. (1994) *The Flesh and the Stone: The Body and the City in Western Civilisation*, London: Faber and Faber.

Sheridan, A. (1977a) 'Translator's note', in J. Lacan, *Écrits: A Selection*, London: Tavistock, vii–xii.
———(1977b) 'Translator's note', in J. Lacan, *The Four Fundamental Concepts of Psycho-analysis*, Harmondsworth: Peregrine Books, 277–282.
Sibley, D. (1981) *Outsiders in Urban Societies*, London: Basil Blackwell.
———(1988) 'Purification of space', *Environment and Planning D: Society and Space* 6: 409–421.
———(1992) 'Outsiders in society and space', in K. Anderson and F. Gale (eds) *Inventing Places: Studies in Cultural Geography*, London: Longman, 107–122.
———(1995a) 'Families and domestic routines: constructing the boundaries of childhood', in S. Pile and N. Thrift (eds) *Mapping the Subject: Geographies of Cultural Transformation*, London: Routledge, 123–137.
———(1995b) *Geographies of Exclusion*, London: Routledge.
Smith, M. P. (1980) *The City and Social Theory*, Oxford: Basil Blackwell.
Smith, N. (1979) 'Geography, science and post-positivist modes of explanation', *Progress in Human Geography* 3: 356–383.
Smith, N. and Katz, C. (1993) 'Grounding metaphor: towards a spatialized politics', in M. Keith and S. Pile (eds) *Place and the Politics of Identity*, London: Routledge, 67–83.
Smith, S. (1990) *The Politics of 'Race' and Residence*, Cambridge: Polity Press.
Snead, J. (1994) *White Screens/Black Images: Hollywood from the Dark Side*, London: Routledge.
Soja, E. (1989) *Postmodern Geographies: The Reassertion of Space in Social Critical Theory*, London: Verso.
Sonnenfeld, J. (1972) 'Geography, perception, and the behavioural environment', in P. W. English and R. C. Mayfield (eds) *Man, Space and Environment*, Oxford: Oxford University Press, 244–250.
Spivak, G. C. (1988) *In Other Worlds: Essays in Cultural Politics*, London: Routledge.
Spivak, G. C. (1993) 'Echo', *New Literary History* 24: 17–43.
Stallybrass, P. and White, A. (1986) *The Politics and Poetics of Transgression*, London: Methuen.
Stea, D. (1973) 'Rats, men and spatial behaviour, all revisited or what animal geographers have to say to human geographers', *Professional Geographer* 25(2): 106–112.
Stea, D. and Downs, R. M. (1970) 'From the outside looking in at the inside looking out', *Environment and Behaviour* 2: 3–12.
Steedman, C. (1995) 'Maps and polar regions: a note on the presentation of childhood subjectivity in fiction of the eighteenth and nineteenth centuries', in S. Pile and N. Thrift (eds) *Mapping the Subject: Geographies of Cultural Transformation*, London: Routledge, 77–92.
Swanson, G. (1995) '"Drunk with the glitter": consuming spaces and sexual geographies', in S. Watson and K. Gibson (eds) *Postmodern Cities and Spaces*, Oxford: Basil Blackwell, 80–98.
Taussig, M. (1993) *Mimesis and Alterity: A Particular History of the Senses*, London: Routledge.
Tester, K. (ed.) (1994a) *The Flâneur*, London: Routledge.
Tester, K. (1994b) 'Introduction', in K. Tester (ed.) *The Flâneur*, London: Routledge, 1–21.
Theweleit, K. (1977) *Male Fantasies, Volume 1: Women, Floods, Bodies, History*, 1987, Cambridge: Polity Press.
———(1978) *Male Fantasies, Volume 2: Male Bodies: Psychoanalyzing the White Terror*, 1989: Cambridge: Polity Press.
Thompson, E. P. (1978) *The Poverty of Theory and Other Essays*, London: Merlin Press.
Thrift, N. (1981) 'Behavioural geography', in N. Wrigley and R. J. Bennett (eds) *Quantitative Geography: A British View*, London: Routledge and Kegan Paul, 352–365.
———(1983a) 'On the determination of social action in space and time', *Environment and Planning D: Society and Space* 1(1): 23–57.
———(1983b) 'Literature, the production of culture and the politics of place', *Antipode* 15: 12–24.
———(1986) 'Little games and big stories: the practice of political personality in the 1945 General Election', in K. Hoggart and E. Kofman (eds) *Politics, Geography and Social Stratification*, Beckenham: Croom Helm, 86–143.
———(1989a) 'Introduction', to new models of civil society in R. Peet and N. Thrift (eds) *New Models in Geography: Volume 2*, London: Unwin Hyman, 149–156.
———(1989b) 'Introduction', to new models of social theory in R. Peet and N. Thrift (eds) *New Models in Geography: Volume 2*, London: Unwin Hyman, 255–266.
Timpanaro, S. (1976) *The Freudian Slip*, London: Verso.

Tolman, E. C. (1946) 'A stimulus-expectancy need-cathexis psychology', *Science* 101: 160–166.
———(1948) 'Cognitive maps in rats and men', in R. M. Downs and D. Stea (eds) *Image and Environment: Cognitive Mapping and Spatial Behaviour*, 1973, Chicago: Aldine, 27–50.
Tuan, Y.-F. (1975) 'Images and mental maps', *Annals of the Association of American Geographers*, 65: 205–213.
———(1976) 'Humanistic geography', *Annals of the Association of American Geographers* 66: 266–276.
———(1977) *Space and Place: The Perspective of Experience*, London: Edward Arnold.
Turner, C. and Carter, E. (1986) 'Political somatics: notes on Klaus Theweleit's *Male Fantasies*', in V. Burgin, J. Donald and C. Kaplan (eds) *Formations of Fantasy*, London: Methuen, 200–213.
Unwin, P. T. H. (1992) *The Place of Geography*, Harlow: Longman.
Waismann, F. (1950–1) 'Analytic-synthetic', *Analysis* 11: 52–56.
Walkowitz, J. (1980) *Prostitution and Victorian Society: Women, Class and the State*, Cambridge: Cambridge University Press.
———(1992) *City of Dreadful Delight: Narratives of Sexual Danger in Late-Victorian London*, London: Virago.
Walmsley, D. J. (1974) 'Positivism and phenomenology in human geography', *Canadian Geographer* 18: 95–107.
Walmsley, D. J. and Lewis, G. J. (1984) *Human Geography: Behavioural Approaches*, London: Longman.
Ward, E. (1984) *Father–Daughter Rape*, London: Women's Press.
Welldon, E. V. (1988) *Mother, Madonna, Whore: The Idealization and Denigration of Motherhood*, London: Free Association Books.
White, G. F. (1945) *Human Adjustment to Floods: A Geographical Approach to the Flood Problem in the United States*, Chicago: Department of Geography Research Paper 29, University of Chicago.
———(1964) *Choice of Adjustments to Floods*, Chicago: Department of Geography Research Paper 93, University of Chicago.
Williams, R. (1990) *Notes on the Underground: An Essay on Technology, Society, and the Imagination*, Cambridge, Massachusetts: MIT Press.
Wilson, B. M. (1980) 'Social space and symbolic interaction', in A. Buttimer and D. Seamon (eds) *The Human Experience of Space and Place*, London: Croom Helm, 135–147.
Wilson, E. (1992) 'The invisible flâneur', *New Left Review*, 199: 90–110.
Witkin, H. A. (1959) 'The perception of the upright', *Scientific American* 200(2): 51–56.
Wolff, J. (1985) 'The invisible flâneuse', *Theory, Culture and Society* 2(3): 37–48.
Wollheim, R. (1991) *Freud* (Second edition), London: Fontana.
Wolpert, J. (1964) 'The decision process in spatial context', *Annals of the Association of American Geographers* 54: 537–558.
———(1965) 'Behavioural aspects of the decision to migrate', *Papers of the Regional Science Association* 15: 159–169.
Wright, E. (1984) *Psychoanalytic Criticism: Theory in Practice*, London: Methuen.
Wright, J. K. (1947) '*Terrae incognitae*: the place of the imagination in geography', *Annals of the Association of American Geographers* 37(1): 1–15.
Young, R. J. C. (1995) *Colonial Desire: Hybridity in Theory, Culture and Race*, London: Routledge.
Zizek, S. (1989) *The Sublime Object of Ideology*, London: Verso.

NAME INDEX

SUBJECT INDEX

THE BODY AND THE CITY

psychoanalysis, space and subjectivity

STEVE PILE

LONDON AND NEW YORK

First published 1996
by Routledge
2 Park Square, Milton Park, Abingdon, Oxon, OX14 4RN

Simultaneously published in the USA and Canada
by Routledge
270 Madison Ave, New York NY 10016

Transferred to Digital Printing 2006

Typeset in Garamond by
Solidus (Bristol) Limited

British Library Cataloguing in Publication Data
A catalogue record for this book is available from the British Library

Library of Congress Cataloguing in Publication Data
The body and the city : psychoanalysis, space, and subjectivity /
Steve Pile.
p. cm.
Includes bibliographical references and index.
1. Psychoanalysis and human geography. 2. Spatial behavior.
3. Personal space. 4. Urban ecology. I. Title.
BF175.4.H84P55 1996
150.19'5–dc20 95–46814

ISBN 0-415-06649-2 (hbk)
ISBN 0-415-14192-3 (pbk)

Publisher's Note

The publisher has gone to great lengths to ensure the quality of this reprint but points out that some imperfections in the original may be apparent

Printed and bound by CPI Antony Rowe, Eastbourne